Liquid Metal Processing

Advances in Metallic Alloys

A series edited by **J.N. Fridlyander**, *All-Russian Institute of Aviation Materials, Moscow, Russia*, and **D.G. Eskin**, *Netherlands Institute for Metals Research, Delft, The Netherlands*

Volume 1

Liquid Metal Processing: Applications to Aluminium Alloy Production
I.G. Brodova, P.S. Popel and G.I. Eskin

This book is part of a series. The publisher will accept continuation orders which may be cancelled at any time and which provide for automatic billing and shipping of each title in the series upon publication. Please write for details.

Liquid Metal Processing

Applications to Aluminium Alloy Production

I.G. Brodova
Institute of Metal Physics, Ekaterinburg, Russia

P.S. Popel
Ural State Pedagogical Institute, Ekaterinburg, Russia

and

G.I. Eskin
All-Russia Institute of Light Alloys, Moscow, Russia

Translated from the Russian by A.I. Kozlenkov

CRC Press
Taylor & Francis Group
Boca Raton London New York

CRC Press is an imprint of the
Taylor & Francis Group, an **informa** business

A TAYLOR & FRANCIS BOOK

CRC Press
Taylor & Francis Group
6000 Broken Sound Parkway NW, Suite 300
Boca Raton, FL 33487-2742

First issued in paperback 2019

© 2002 by Taylor & Francis Group, LLC
CRC Press is an imprint of Taylor & Francis Group, an Informa business

No claim to original U.S. Government works

ISBN-13: 978-0-415-27233-9 (hbk)
ISBN-13: 978-0-367-39685-5 (pbk)

Visit the Taylor & Francis Web site at
http://www.taylorandfrancis.com

and the CRC Press Web site at
http://www.crcpress.com

This book has been produced from camera-ready copy supplied by the authors

British Library Cataloguing in Publication Data
A catalogue record for this book is available from the British Library

Library of Congress Cataloging in Publication Data
A catalog record for this book has been requested

Contents

Preface

The overwhelming majority of technological processes for producing metallic alloys involve the conversion of initial materials into the molten condition and the subsequent solidification of the system at various, sometimes very high, cooling rates. In an attempt to improve the structure and service properties of ingots, castings and deformed semi-finished products, process engineers have paid considerable attention to the search for optimal solidification conditions. The regimes of heat and mechanical treatment of cast metal are progressively improving. Only the first stage of the process – initial melt – is traditionally of marginal interest to metallurgists. In most cases, attempts to affect the system at this stage are confined to additional alloying with the aim of optimizing the system composition, and to refining the system in an effort to remove deleterious impurities.

At the same time a great body of data has been accumulated in scientific periodicals over the past 30–40 years, which indicates that metallic melts are dynamical systems of high complexity. They can exist in different structural states and make transitions between these states under the influence of various external actions. The role of the structural state of the initial melt in the formation of the structure and properties of ingots produced from this melt and, consequently, the structure and properties of deformed semi-finished products is also established. As applied to steels, cast iron and nickel alloys, these facts are systematized in the collective monograph under the title *Liquid Steel* * published in the USSR in 1984. However, information regarding aluminium alloys has not been generalized to date, although the data accumulated in this field are equally impressive.

In the authors' opinion, this monograph may fill this gap. It owes its origin to the collaboration of three researchers who over the past 20–30 years have studied various aspects of interrelation between the structure of aluminium melts and the structure and properties of ingots formed in solidification of these melts, or have searched actively for effective ways of influencing liquid metals to improve the quality of solid metals. We consider in this monograph a few possible ways of governing the structure and properties of cast metal by applying different external actions to the initial melt. In Chapter 1 the possibilities of realization of various structural rearrangements in liquid metals and alloys

---

* Baum, B.A., Hasin, G.A., Tyagunov, G.V., *et al.*, (1984) *Zhidkaya stal'* (Liquid Steel), Moscow: Metallurgiya

are considered, and the most promising methods for their implementation in industrial conditions are separated out, namely, overheating of melts above the liquidus up to specific temperatures for each composition [the so-called heat–time treatment (HTT) of melts] and the modification of melts by introducing additions of various kinds. Chapter 2 is devoted to the study of the influence of HTT of liquid aluminium alloys on the structure and properties of ingots. Chapter 3 contains material concerning modification of aluminium alloys, and in Chapter 4 we consider the theory and practice of ultrasonic treatment (UST) of liquid metal as an alternative to HTT of melt and to its modification.

The Introduction and the first chapter were written by Professor P.S. Popel (Ural State Pedagogical University); the second and third chapters, by Professor I.G. Brodova (Institute of Metal Physics of the Ural Division of the Russian Academy of Sciences); and the final, fourth chapter, by Professor G.I. Eskin (All-Russia Institute of Light Alloys, Moscow), with the exception of Sections 2.1, 3.4.1, 4.2.1, and 4.2.3, which were written by P.S. Popel.

Naturally, a significant part of the book is based on original material obtained by the authors. We also attempted to analyze the data of other authors and would like to present our apologies to those whose results are not included in this book or were not properly considered due to inevitable subjectivity in the selection of material.

This book is intended, above all, for specialists involved in the production of aluminium alloys and articles produced from these alloys. We hope that it will also be of interest for scientists involved in theoretical and experimental studies of liquid metallic solutions and solidification processes.

The authors would like to express their sincere gratitude to N.P. Kurochkina and D.V. Bashlykov for technical assistance in preparing the Russian version of the manuscript.

List of Abbreviations

DMA	master alloy obtained by casting in rolls
EPMA	electron probe microanalysis
EXAFS	extended X-ray absorption fine structure (spectroscopy)
FIM	field ion microscopy
HMA	master alloy obtained by rapid quenching of melt
HREM	high-resolution electron microscopy
HTT	heat–time treatment (of melts)
OMA	chill cast master alloy obtained from overheated melt
SAD	selected area diffraction (microdiffraction)
SADP	selected area diffraction pattern
SANS	small-angle neutron scattering
SAXS	small-angle X-ray scattering
SMA	pig cast master alloy
UST	ultrasonic treatment

List of Symbols

A	surface area of disperse particle
a	crystal lattice parameter
B	accumulation factor
b	halfthickness of transition layer at the boundary of disperse particle
C_i	concentration of the ith component, wt. %
c	speed of light in vacuum
D_{gr}	average cast grain size
d	density
E	energy of the virtual bound state relative to the bottom of the conduction band
e	electron charge
F	Helmholtz free energy
f_z	force projection on the direction of magnetic field strength
G	Gibbs free energy
G_v	free energy of activation of viscous stream
$g(r)$	atomic radial distribution function
$g_{ig}(r)$	partial function of atomic radial distribution
H	magnetic field strength
H_μ	microhardness
HB	Brinnel hardness number
ΔH	integral heat of mixing
ΔH_i	partial heat of mixing of the ith component
h	sample height
I	moment of inertia
i	electric current
J	intensity of radiation beam
K_0	distribution coefficient
K_D	diffusion coefficient
K_l	coefficient of torsion
K_m	modification coefficient
K_{str}	structure misfit coefficient
k	Boltzmann constant
k_F	Fermi momentum

L	average linear size of crystal
l	path length of gamma-ray photons in sample
M	angular momentum
m	mass
N	number of phase precipitates per unit of area
N_a	Avogadro number
N_c	number of atom in cluster
N_{cel}	number of atoms in unit cell
N_d	number of d electrons
N_i	number of atoms of kind i
$N(\varepsilon)$	density of electronic states
n	atomic number density
n_c	number of clusters in unit volume
[O]	oxygen concentration, wt. %
P	power of ultrasonic oscillations
P_m	magnetic moment of sample
p	pressure
p_m	specific magnetization
Q	surface activity of solute
q	screening length
R	radius
R_c	cluster radius
R_g	gas constant
R_H	Hall constant
R_φ	interatomic potential range
S	entropy
$S(k)$	structure factor
$S_{ig}(k)$	partial structure factor
s	scattering cross-section
T	temperature
T_{an}	temperature of anomalous point at the temperature dependence of the property
T_c	Curie temperature
T_{c1} and T_{c2}	temperatures at which the deviation of temperature dependences of the property obtained in heating and subsequent cooling of the melt appears and stops to increase, respectively
T_c	casting temperature
T_e	eutectic temperature
T_{hom}	melt homogenization temperature
T_L	liquidus temperature
T_m	melting temperature
ΔT	melt overheating above liquidus line
δT	melt supercooling at the solidification front
t	time
τ_{cr}	time elapsed from master alloy introduction in the melt to the appearance of modifying effect (incubation period)

U	internal energy
v	volume
v_μ	molar volume
v_i	partial molar volume of the ith component
V	cooling rate
V_s	ultrasound speed
$W\gamma$	energy of gamma-ray photons
x	the component concentration in atomic fractions or at.%
x_e	eutectic concentration
Z	atomic number
z_i	coordination number for the ith coordination sphere
α	thermal expansion coefficient
Δ	logarithmic decrement of damping of oscillations
δ	tensile elongation
ε_F	Fermi energy
χ	magnetic susceptibility
η	dynamical viscosity coefficient
Φ	volume fraction of disperse particles or clusters
φ	angle of twist of suspension thread
$\varphi, \%$	relative tensile contraction
Λ	electrical conductivity
λ	wavelength
μ	mass gamma-ray attenuation factor
μ_i	chemical potential of the ith component
ν	kinematic viscosity coefficient
ν_i	number of moles of the ith component
θ	scattering angle
ρ	electrical resistivity
σ	surface or interphase tension
σ_u	ultimate tensile strength
$\sigma_{0.2}$	nominal yield point
τ	period of oscillations
$\varsigma\cdot$	angular rate
ω	interchange energy per one atom
ω_μ	interchange energy per one mole of solution
ξ	filling factor
ζ	parameter taking into account the effect of crucible bottom on damping of oscillations

Introduction

In modern theories of solidification of metals and alloys the initial melt is considered to be a continuous medium with uniform and perfectly disordered atomic distribution. As a crystal grows, the liquid phase supplies the "construction material" to it in the form of individual atoms. This concept is based on the idea of liquid as a strongly compressed gas with chaotic distribution of atoms and their Brownian motion.

However, as early as the 1920s the concepts of microheterogeneity of metallic melts had begun to form. A certain resemblance between X-ray scattering curves obtained from the melt and from the initial crystalline specimen led Stewart to the idea of the existence in the liquid phase of groups of atoms or molecules having a certain mutual coordination (short-range order) but retaining their mobility. The formation of such groups is due to the same interparticle forces that cause the system to solidify at lower temperatures. For this reason, the hypothesis put forward by Danilov [1] about the close similarity between the structure of such formations and the crystal structure was accepted as quite natural. This opinion was strengthened after the issue of Frenkel's monograph [2] in which, in contrast to the commonly accepted view, the liquid was treated as a solid with "elements of disorder" rather than as a uniform compressed gas. Along with the ordering in the arrangement of particles, the fundamentally oscillatory character of their motion was substantiated. Successful application of these concepts by Frenkel to the analysis of various processes in the liquid phase (in particular, to the treatment of diffusion and viscosity) contributed to a wide dissemination of the "quasi-crystalline" concept on the structure of liquid metals and alloys.

Stewart–Danilov's model was further developed by Mott and Herney [3], who used it to construct the theory of fusion, and by H. Eiring et al. [4], who allowed for the distinction between the structure of ordered regions and the structure of the initial solid body. One of the most consistent proponents of Frenkel's concepts is Ubbelohde [5, 6]. Based on the nuclear magnetic resonance (NMR) studies of the structure of liquid metals, which point to an insignificant change of short-range order in fusion, Ubbelohde separates three regions in a liquid phase: (1) the region with the structure of the initial crystal (crystalline clusters); (2) the region with a structure different from that of crystalline clusters, but also ordered (anticrystalline clusters); and (3) the disordered zone. Clusters in liquid metals containing about 100 ator-5 (the estimate is based on neutron small-angle scattering data) are treated by Ubbelohde as dynamical structures of fluctuation origin, whose lifetime ($\sim10^{-10}-10^{-11}$ s) is a few orders of magnitude longer than the time of settled life of atoms ($\sim10^{-13}-10^{-14}$ s). In Ubbelohde's opinion, the tendency of melts to supercooling is determined by the ratio of numbers of crystalline and anticrystalline clusters.

Attempts were made repeatedly to estimate the size and fraction of clusters in molten metals. Dutchak and Klym [7] inferred from analysis of diffraction patterns the cluster radii in the range from 1.9 to 2.2 nm. Regel and Glazov [8] used Einstein's formula for the suspension of solid spherical particles to interpret the temperature curves of viscosity η. The volume fraction of clusters near the melting point derived from the deviation of the $\eta(T)$ curves above the melting point from the exponential dependence was estimated to be 1–10%. Ershov and Poznyak [9] determined cluster radii using the cluster mechanism of crystal fusion proposed by them, and obtained values in the range from 0.46 nm for lead to 2.3 nm for iron. These radii were used to calculate the volume change on melting and viscosity for several metals and reasonable agreement with experiment was obtained.

Despite certain progress achieved by applying the quasi-crystalline approach to the structure of liquid metals, the idea about the absence of consistency between the structures of the liquid and solid phases has become increasingly popular since the early 1960s. The basis for this view was provided by the works of Bernal on the geometric modelling of disordered structures [10] and by successful application of a very simple model of rigid spheres to the calculation of properties of many liquid metals [3, 11]. Reasonable agreement between the calculated characteristics and experimental data was achieved by varying only two model parameters of spheres – the radius and the filling factor. Such models cannot be considered as rejecting the existence of microheterogeneity even in one-component melts because in Bernal's model atomic configurations are also present with different probabilities [12]. However, in most cases these configurations are specific for the liquid state and do not occur in the solid state. In particular, they include groups with elements of 5-fold symmetry which is "forbidden", as was thought previously, for crystalline bodies.

Although in the 60 and 70s the two above-mentioned views of the structure of short-range order in liquid metals were considered as alternative to each other, these concepts appeared to draw closer together later due to the results of theoretical works, diffraction and computer experiments, and new data obtained from measurements of various properties of metallic melts. In particular, much evidence was obtained that atomic groupings of the icosahedron type with 5-fold symmetry dominate in the structure of most one-component metallic melts [12] at sufficiently high overheating above the melting point. However, the icosahedron cannot be conceived as a structure unrelated to the initial crystal because such a configuration can be obtained by a set of minimal atomic displacements from the cuboctahedron which forms the nearest environment of particles in the face-centered cubic (fcc) lattice [13]. On the other hand, rather elongated ($\sim1-10$ mm) quasicrystals of 5-fold symmetry were revealed [14] in several rapidly cooled alloys (for example, in Al_4Mn).

Despite the different views of the nature of short-range order in metallic melts, most researchers acknowledge the existence in melts of some kind of atomic clusters, i.e. the inhomogeneity of melts on a microscopic level. However, the character of structural transformations occurring in melts at elevated temperatures, or under the action of different external factors remains disputable. Traditional views reduced to the idea that an increase in temperature gives rise to the progressive smearing of the short-range atomic order by thermal motion, so that the atomic distribution becomes perfectly disordered at sufficiently high temperatures. In the course of subsequent cooling, progressive ordering of the system occurs, increasing as the crystallization point is approached. As was mentioned above, the lifetime of clusters in liquid metals is of the order of $10^{-10}-10^{-11}$ s. For this reason, at cooling rates achievable by a known technological processes (up to 10^6-10^7 K/s) the melt at any time has a short-range order structure that is in equilibrium at the given temperature. Therefore, as the solidification point is approached, the melt "forgets" its past history (heating to more or less high temperatures and other external effects) and at any time has a definite structure. Under such conditions, a change in the technological melting regime should not affect the process of solidification of the alloy at moderate cooling rates or the course of amorphization on rapid quenching, and therefore, the structure of the solid metal and its properties should also remain unchanged.

Precision structural studies and measurements of properties of metallic melts have shown that the real processes occurring in the liquid phase are much more intricate.

First of all, a variety of anomalies were found in the temperature dependences of properties of liquid metals, which pointed to significant changes in their structure occurring in narrow temperature ranges. The fact that these anomalies are similar to the same features in the temperature curves of the properties of crystalline metals near the phase-transition points, led the researchers to the assumption that abrupt changes are possible in the short-range order structure of melts. Since most of the known phase transitions are preceded by overheating or supercooling of the initial phase, the assumption was made that in liquid metals a new structure of short-range order also appears, not immediately after passing the point of the assumed transformation, but only after going deeper into the region of its existence. In this case it becomes possible to heat the melt above the temperature of a given anomaly, thereby converting the melt to a new (high-temperature) structural state, and then to allow the melt to cool down rapidly to the solidification point in such a way that the inverse transformation would have no time to occur. There are definite reasons to think that in this case solidification will differ from the standard process and the structure of the formed ingot will be different.

Much more diversified possibilities of governing the structure and the properties of cast metal by means of various actions on the initial melt appear in cases when the system consists of two or more different components, i.e. if it is an alloy.

The introduction of even tiny amounts of the second component ($\sim 0.01-0.1$ at. %) is accompanied by the formation of a new type of clusters surrounding impurity atoms. In contrast to clusters formed in pure metals, these impurity clusters, containing from 100 to 1000 atoms, are stable complexes existing for unlimited time in a fairly wide temperature range above the liquidus point. During solidification of the melt, some of these clusters may play the role of nuclei of the new phase, i.e. they may actively influence the solidification process. At the same time, the size and the number of impurity clusters in the melt depend on the type of the introduced additive and on its concentration. By varying the impurity elements and their content, one can effectively influence the structure of castings and semi-finished products. We should note immediately that this process differs fundamentally from traditional processes of modification of cast metal, in which an additive is introduced with the aim of changing the conditions of growth of the crystals already existing. In our case, the additive modifies the structure of the initial melt and the conditions for nucleation of the solid phase in the melt.

If the concentration of the second component is of the order of 1 at. % or more, the interaction between impurity clusters leads to their destruction. However, the melt does not become homogeneous on the microscopic level. A certain ordering is observed in the configuration of neighbouring atoms of different sorts, which is called the chemical short-range order. If the interaction energy of like atoms exceeds the energy of interaction between unlike atoms, there arise groupings consisting chiefly of atoms of the same component. In the opposite case, each atom tends to have like atoms as the nearest neighbours. As the content of the second component increases, the melt passes through a sequence of changes in the chemical ordering of its atoms, remaining macroscopically homogeneous. These changes occur in comparatively narrow concentration ranges and the structure of the casting obtained from the melt depends on the degree of completion of these changes. Therefore, by varying the content of the alloy components even within the limits admissible by compositional standards we can appreciably affect the quality of cast metal, and the reason for this influence lies in the change of chemical short-range ordering in the initial melt.

The most significant evidence of microheterogeneity of melts is in the eutectic-type systems to which most of the casting aluminium alloys pertain, and in systems with a miscibility gap in the liquid state outside this region. In both cases diffraction experiments reveal the existence of microgroups enriched with like atoms. The characteristic scale of this microheterogeneity, according to various estimates, is of the order of 10 nm, i.e. it is far in excess of the short-range ordering scale in melts. The results obtained over the past 15 years show that the microheterogeneous state in liquid eutectics and in systems with segregation is not thermodynamically stable; it is, in essence, a metastable state with a near-liquidus lifetime of the order of tens of hours or, perhaps, a nonequilibrium state with a relaxation time of the order of 1 to 10 hours. This state is inherited from heterogeneous charge materials or from the segregated initial melt retaining, therefore, certain "memory" of their features. As a result of heating or other external actions on the melt, such microheterogeneity may be irreversibly destroyed. Then, on further cooling at any rate, the system will approach the onset of solidification or the immiscibility dome in a structural state which is considerably different from that which it had upon melting or passing through the dome in heating. The scale of the initial microheterogeneity is close to the characteristic radius of critical nuclei of the new phase. Therefore, no energy hindrances arise for the onset of the phase transition in cooling the melt existing in the microhetero-geneous state. On the other hand, once the microheterogeneity has been destroyed, the melt does not contain ready centres of the new phase and its solidification or separation into two immiscible liquids begins only after deep supercooling and proceeds in a substantially different regime. As a result, an ingot forms with structure and properties different from those realized upon solidification of the microheterogeneous melt.

As a consequence, we have a few possibilities for controlling the structure and properties of cast metal by means of various external actions on the initial melt, which will be considered further in this monograph.

To complete this brief introduction to the topics of this book, we note that the concepts of the character of structural rearrangements in melts, which were discussed above and will be considered in detail in Chapter 1, are to some extent hypothetical. They rest on experimental data obtained chiefly by measuring their properties indirectly related to the state of atomic and electron subsystems rather than from direct structural studies of melts. Unfortunately, to date metallic melts remain "opaque" objects for researchers. This is related both to the lack of reliable methods for structural analysis of liquid metals and to the inadequate theoretical description of such objects. However, the mere facts of structural changes occurring in melts at definite temperatures, or under the influence of various external actions are, in our opinion, beyond question. The prospect of using these effects for improving the technology of melting and casting aluminium alloys is also reliable.

Chapter 1

Structural Rearrangements in Metallic Melts

1.1 Experimental Methods for Investigating Structural Rearrangements in Melts

Most objective information on the structure of short-range order in liquid metals can be derived from diffraction studies with the use of X-ray or synchrotron radiation, and electron and neutron beams [1]. These studies provide the possibility of experimentally determining the so-called structure factor $S(k)$ of the melt, the Fourier transform of which gives the radial atomic distribution function $g(r)$ (Fig. 1.1). The $g(r)$ function has a sense of the density of probability to find an atom of the melt at a distance r from the given atom, i.e. it characterizes the time-averaged short-range order in the atomic arrangement.

However, the interpretation of experimental diffraction patterns proves to be ambiguous even in cases when the simplest two-component liquid alloys are studied. In this case, it is necessary for the characterization of short-range ordering to determine three radial distribution functions $g_{ij}(r)$, each being the density of probability of finding an atom of kind j at a distance r from an arbitrarily chosen atom of kind i. These functions can be found by the Fourier transformation of the so-called partial structure factors $S_{ij}(k)$. The problem lies in the fact that the determination of these factors from experimentally obtained scattering curves requires three independent diffraction experiments with different radiations, or different isotopes of one of the alloy components. Such experiments are extremely laborious and are being conducted presently only episodically at the best-equipped scientific centres. In an attempt to circumvent this obstacle, the researchers have to make inadequately reliable assumptions when analyzing diffraction patterns. As a result, identical diffraction curves are treated by different authors from different positions. For multicomponent alloys, the interpretation of the results of diffraction experiments is virtually impossible.

A wide use of diffraction measurements for detecting structural rearrangements in metallic melts is also hindered by the relatively long duration of these experiments. In particular, the reliable recording of an X-ray scattering curve usually takes a few hours. This does not permit the determination with required accuracy of temperatures and

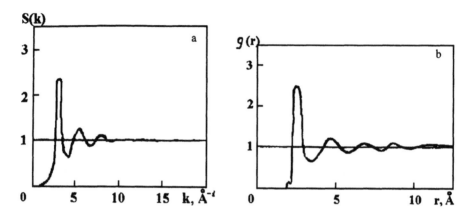

Figure 1.1   (a) Structure factor $S(k)$ and (b) radial atomic distribution function $g(r)$ of liquid iron at 1560°C.

concentrations at which significant rearrangements occur in the structure of melts, as well as the study of various relaxation processes involved in such rearrangements. Electron diffraction experiments take a much shorter time but their use is limited in some cases by high volatility and chemically aggressive properties of melts. The maximum efficiency is attained by using synchrotron radiation [2], which, in addition, allows studies with the quite promising EXAFS method [3], which significantly facilitates the derivation of partial structure factors. However, these methods have begun to develop in the past decades and have thus far been inadequately developed in theoretical and instrumental respects.

For this reason, a great body of information on the structural rearrangements occurring in liquid metallic solutions under the influence of temperature and compositional changes or various external actions is obtained as a result of the study of their properties sensitive to a greater or lesser degree to such rearrangements (these properties are often called structure-sensitive). In these experiments various anomalies on the temperature and concentration curves of the measured properties or their significant response to the external factors acting upon the melt are recorded. These data enable one to draw the conclusion that structural transformations exist. Comparing the behavior of various properties of the melt near the transformation point, we can gain a tentative impression of its physical nature. At the final stage, the resulting pattern is augmented by the results of diffraction studies and theoretical analysis.

Experimental methods for measuring the properties of high-temperature, often chemically aggressive metallic melts, which are used for solving the above problems, must be, wherever possible, contactless, have rather high sensitivity to small variations of the measured property, allow its permanent monitoring with an immediate display of the results obtained, and must be highly efficient and not labor-intensive. In addition, it is quite desirable to have at one's disposal at least a few methods allowing the measurements of the properties of the sample studied both in the liquid and in the solid state, as well as the possibility to measure these properties in the course of solidification. This makes it possible to assess to what extent the detected change in the melt structure affects the process of melt solidification and the properties of the ingot being formed.

The method of determination of density by measuring attenuation of the beam of gamma-ray photons in the specimen under study satisfies the above requirements to the greatest extent. It is based on the known law of attenuation of the beam of gamma-ray photons in the material with density d:

$$J = BJ_0 \exp(-\mu l d),$$ (1.1)

where J_0 and J are the beam intensities upstream and downstream the absorber, respectively; l is the path length of gamma-ray photons in the specimen; μ is the mass attenuation coefficient depending on the chemical composition of the specimen and the energy W_γ of gamma-ray photons; and B is the accumulation factor depending on the setup geometry and W_γ. By measuring J_0, J, l and calculating B, we can determine the specimen density using the known attenuation coefficient μ:

$$d = (\ln(J_0/J) + \ln B)/(\mu l).$$ (1.2)

Precision measurements of the density of melts are carried out, as a rule, by using the relative variant of the gamma-ray absorption technique in which attenuation of the beam of gamma-ray photons by the melt under investigation is compared to the attenuation of gamma-rays in a standard solid specimen of the same composition and known density [4]. In this variant of the method, the mass attenuation coefficient μ does not appear in the calculation formula and systematical errors in the main experiment and in that performed on standard specimens compensate for each other.

However, the relative method has significant disadvantages too. Practice shows that the chemical composition and density along the path of a narrow gamma-ray beam may differ significantly from their mean values for the standard specimen as a whole. Due to such inhomogeneities in the standard specimen, uncontrolled errors might be introduced into the results of melt density measurements. Moreover, precision measurements of concentration dependences of the density of melts in a wide compositional range with a small concentration step are required to gain information on variations in the melt structure caused by alloying elements. This problem can be successfully solved only on condition that the alloying elements are introduced into the sample directly in the course of experiment. The use of the relative version of the gamma-ray absorption method does not provide such a possibility because significant variations in the melt composition give rise to a considerable difference between the attenuation coefficient μ of the melt and that of the standard specimen.

Most investigators, who have used the absolute version of this method, have seen the major impediment in the necessity of invoking the reference values of the mass attenuation coefficient μ. The number of such reference sources is limited and the accuracy of available data is rather low. However, an analysis of the literature data performed in [5] shows that when the ^{137}Cs source with the gamma-ray energy $W_\gamma = 0.6616$ MeV, which is most commonly used in gamma-ray densitometry, is employed for elements with moderately high atomic numbers (with $Z < 30$), the μ values can be calculated with an accuracy of better than 0.15% and, therefore, no fundamental impediments exist to the measurement of

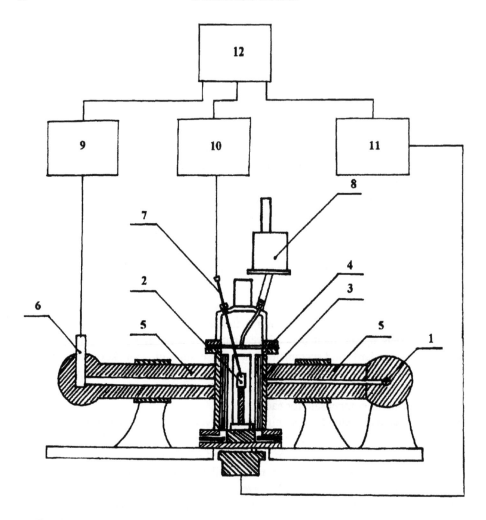

Figure 1.2 Experimental setup for measuring density of high-temperature melts: (1) gamma-ray radiation source, (2) crucible filled with melt, (3) water-cooled housing of the vacuum resistance furnace, (4) molybdenum heater, (5) collimators, (6) gamma-ray scintillation detector, (7) tungsten–rhenium thermocouple, (8) metering tank for introducing alloying additives into melt, (9) gamma-ray beam intensity measuring unit, (10) melt temperature measuring unit, (11) heater controller, (12) computer.

the melt density by the absolute gamma-ray absorption method with an accuracy of the same order.

The construction of one of the most suitable gamma-ray densitometers is shown in Fig. 1.2. The narrow gamma-ray beam from the ^{137}Cs source formed by the collimator passes in the radial direction through the cylindrical crucible containing the specimen under investigation. The crucible is placed in a vacuum furnace with a molybdenum heater, which enables one to keep the temperature as high as 1900°C and to conduct measurements in vacuum at pressure of about 1 Pa or in the inert gas atmosphere. The temperature is measured by a thermocouple with a protecting tube, which is immersed in the melt to be studied. The intensity of the beam which passed through the specimen is recorded by the scaler unit connected with the output of the scintillation gamma-ray detector. The initial intensity I_0 of the beam is determined in the calibration experiment in which the specimen is replaced by an empty crucible. The setup operation control and data processing are accomplished by a computer.

The sensitivity of the gamma-ray absorption technique to variations in the melt density, which reflect the structural rearrangements occurring in the melt, can be improved to the value of order 0.01% by the proper choice of the optimal activity of the source and the exposure in measurements of the intensities J_0 and J. Up to 1000 experimental points are obtained in an individual experiment. This enables one to employ rather sophisticated methods of statistical processing and to reliably detect various anomalies in the temperature and concentration curves of density, as well as to study relaxation processes in melts. The gamma-ray absorption method allows measurements in the course of solidification of the specimen and afterwards in the solid state, i.e. enables one to monitor the influence of the structural rearrangement of a liquid metal on the formation of the properties of the ingot.

Viscosity of liquids, as well as density, is related to the structure of the short-range order and depends on the interaction between particles. Viscosity is also quite sensitive to the presence of disperse inclusions in the melt. Their dimensions and the volume content in the melt can be calculated from experimental viscosity values using a series of theoretical dependences [6]. As a consequence, viscosity measurements may give information on structural rearrangements in the system under investigation.

The above-mentioned requirements for the methods of investigating the properties of high-temperature melts restrict significantly the use of conventional stationary methods of measuring viscosity of liquids. At present the determination of the kinematic viscosity ν from the rate of damping of torsion oscillations of the cylindrical crucible filled with the melt under study is considered to be the most reliable and well-approved method. (The kinematic viscosity ν of a liquid is related to its dynamical viscosity η by the simple relation $\nu = \eta/d$, where d is the density.) The crucible is suspended on an elastic thread and its torsion oscillations begin after the initial twisting of the suspension thread, which is usually initiated by a short turn-on of the rotating magnetic field. The logarithmic decrement Δ and their period τ are measured. Next, using the solution of the corresponding hydrodynamical problem relating these parameters with the kinematic viscosity ν of the melt, the latter quantity is determined.

E.G. Shvidkovskii was one of the first researchers who solved the above hydrodynamical problem [7]. By introducing a series of simplifying assumptions, he derived the formula for calculating the kinematic viscosity of low-viscous liquids from the observed parameters of torsion oscillations of the system:

$$\nu = (1/\pi)(I/mR)^2(\Delta - \tau\Delta_0/\tau_0)^2/(\tau\zeta^2). \tag{1.3}$$

Here, m is the mass of the liquid; I is the moment of inertia of the system; Δ and Δ_0 are the logarithmic decrements of damping of oscillations in the systems with and without liquid, respectively; τ and τ_0, respective oscillation periods; R, the radius of the crucible, and ζ, the parameter taking into account the influence of the crucible bottom on damping.

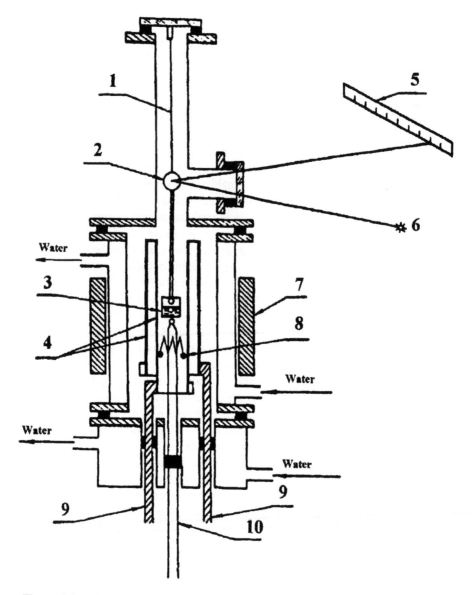

Figure 1.3 Experimental setup for measuring viscosity by the damping of torsion oscillations of a crucible filled with melt: (1) pendulum thread, (2) mirror for detecting oscillations, (3) crucible with specimen, (4) main graphite heater, (5) oscillating amplitude scale, (6) source of narrow light beam, (7) winding for creating magnetic field, (8) additional heater, (9) current supply for the main heater, (10) tungsten–rhenium thermo-couple.

To determine ζ, the expression

$$\zeta = 1 - 3x/2 - 3x^2/8 - a + 2R(b - cx)h, \tag{1.4}$$

was proposed in [7]. Here $x = \Delta/2\pi$; a, b, and c are the coefficients; and h is the height of metal in the crucible. The values of a, b, and c are determined from tables or graphs as a function of the main parameter of the problem: $y = 2\pi R^2/\tau v$.

To calculate the kinematic viscosity, it is necessary to determine the periods and the logarithmic decrements of damping of the pendulum oscillating system with and without liquid. The axial moment of inertia is determined from two series of observations of the period of oscillations: with the empty crucible and when the crucible is charged with the standard whose moment of inertia I_{st} is known (τ_0 and τ_{st}, respectively). The viscosity is calculated from formulas (1.3) and (1.4) by the iterative procedure. First, the viscosity value is calculated from (1.3) with $\zeta = 1$. Then, the coefficients a, b, and c are found from tables or graphs for the relevant y value, which makes possible the calculation of ζ from (1.4). If ζ differs greatly from unity, it is used for determining the second approximation for v. Calculations are repeated until the ζ value calculated in a certain cycle coincides with its value in the preceding cycle within the required accuracy.

With the wide advent of computers into laboratory practice, the need for a series of simplifications in solving the hydrodynamic problem underlying the given method became redundant. At the present time, the theory of torsion viscosimeters is most completely developed by V.P. Beskachko. One of the most appropriate constructions of this instrument is shown in Fig. 1.3.

The method of measuring viscosity from the decrement of torsion oscillations of a crucible with the melt to be studied is an absolute method because it does not call for the use of the standard liquid with a known v value for calibration. Also, there is no need to take into account the distorting effect of the surface tension of the melt and the crucible wetting. The method allows measurements in a wide range of viscosity values with an accuracy of better than 4–5%. The sensitivity of the method to viscosity variations caused by structural rearrangements in the melt may be as high as 1%. Owing to these advantages, this method has found a wide application in viscosimetry of high-temperature metallic melts.

Measurements of viscosity of melts by the above method can be combined in a single setup with measurements of the *electrical resistivity* ρ (or the *electrical conductivity* $\Lambda = 1/\rho$). This characteristic is related to the state of the electron subsystem of the melt. However, the scattering of free electrons by ions is determined by the structure factor $S(k)$ or by a set of partial structure factors $S_{ij}(k)$. For this reason, the electrical resistance of the melt may also be treated as a property sensitive to variations in the atomic ordering.

The determination of ρ from the angle of twist of the specimen suspended on the elastic thread under the influence of the rotating magnetic field is the most commonly used contactless method of measuring electrical resistivity of liquid metals and alloys.

The most comprehensive theoretical justification of this method was given by Regel [8] who considered the behaviour of the conducting sphere of radius R in the rotating magnetic field and calculated the moment of forces M acting upon this sphere:

$$M = 2\pi\Lambda\Omega H^2 R^5/15c^2. \tag{1.5}$$

Here Ω is the angular rate of rotation of the magnetic field with strength H and c is the speed of light.

Expression (1.5) is correct in the case when we can neglect the effect of self-induction on the M value, which is valid if

$$R(2\pi\Omega\Lambda/c)^{1/2} \ll 1. \tag{1.6}$$

This condition is satisfied for sufficiently small R, i.e. for small-sized specimens (usually less than 1 cm^3).

Given the coefficient of torsion K_1 of the suspension thread, we obtain the equation relating the angle φ of twist of suspension thread to the conductivity Λ of the melt by equating the moment of the elastic force and the torque of the magnetic force at the point where these moments compensate each other.

The relative version of this method is used, as a rule, for the determination of Λ. It is based on the proportionality between the specimen conductivity and the angle φ of thread twisting. Indeed, combining the constants in (1.5) and bearing in mind the relationship between the radius and the volume of the specimen, $R^5 \sim v^{5/3}$, we obtain the equation $M = K_2\Lambda v^{5/3}$ in which K_2 is the resulting proportionality factor. Then, with the equal moments, we obtain

$$\Lambda = K_1\varphi/K_2 v^{5/3}, \tag{1.7}$$

and the formula for resistivity takes the form

$$\rho = K v^{5/3}/\varphi, \tag{1.8}$$

where K is the instrumental constant, which is determined in special experiments with metals of known conductivity.

Thus, the measurement of resistivity reduces to the determination of the stationary angle of twisting of the suspension system with the specimen under study. Regel showed that the proportionality between Λ and φ takes place also for cylindrical specimens if their radius R is comparable to the height h. In this case the equation for ρ takes the form

$$\rho = (Ki^2hR^4)/\varphi, \tag{1.9}$$

where i is the current generating the rotating magnetic field.

The experimental setups implementing the above method of measuring resistivity of melts do not differ fundamentally from the viscosimeter shown in Fig. 1.3. As was indicated above, there is a possibility of combining viscosity and conductance measurements in a single instrument. However, it should be borne in mind that the theory of viscosimetric method is valid at $h \gg R$, whereas the calculation of resistivity from formula (1.9) is justified for commensurate R and h values. There exists only a fairly narrow region of measured values where parallel determinations of v and ρ for the same specimen are possible. In most cases the researchers are forced either to change the suspension system in passing from viscosity to conductance measurements or to measure these properties of the melt with separate setups, which usually differ only in the suspension system's parameters.

The method for measuring resistivity in the rotating magnetic field cannot be regarded as rigorously well-founded. When deriving the formulas of this method, the effect of specimen viscosity on the magnitude of the moment of force acting upon the specimen is ignored. For specimens of cylindrical shape, formula (1.9) gives only approximate values, and the error introduced by approximations cannot be correctly estimated. As a consequence, the estimates of accuracy of this method (from 1 to 5%) presented in some works seem to be unjustifiably optimistic. Nevertheless, the use of this method as a sensitive indicator of structural rearrangements in the melt proved to be fruitful.

Another property which is strongly responsive to the changes in the state of atomic and electron subsystems of the melt is its *magnetic susceptibility* χ. The study of susceptibility gives, in particular, information on the distribution of electron density in the specimen and on the effective magnetic moment per atom of the metal. Both these characteristics depend in turn on the structure of the short-range order in the melt.

In high-temperature measurements of magnetic susceptibility of weak magnetics, such as liquid metals and alloys, the so-called ponderomotive methods have gained the widest acceptance. These methods are based on the measurement of the force acting upon the specimen placed in the non-uniform magnetic field:

$$f_z = P_m(\partial H/\partial z), \qquad (1.10)$$

where f_z is the force projection onto the direction of the field strength gradient $\partial H/\partial z$ and P_m is the magnetic moment of the specimen. Since $P_m = p_m m = m\chi H$ (m is the specimen mass and p_m is the specific magnetization), we have

$$f_z = m\chi H(\partial H/\partial z). \qquad (1.11)$$

Measuring the force f_z, we can determine, given the product $H(\partial H/\partial z)$, the magnetic susceptibility of the specimen.

In the instruments employing Faraday's method, the crucible with the melt is positioned directly in the region of the maximum field gradient; the size of the specimen being chosen so as to prevent significant changes in the $H(\partial H/\partial z)$ value.

On account of difficulties related to the need for exact fixation of the specimen position and the use of polepieces of special shape, the relative variant of Faraday's method is typically employed, in which a comparison is made of the f_z and f_{z0} values obtained, respectively, in experiments with the melt of interest and the standard specimen with the known magnetic susceptibility χ_0. Then we have

$$\chi = (m/m_0)\chi_0(f_z/f_{z0}), \qquad (1.12)$$

where m and m_0 are the masses of the specimen and the standard, respectively.

In instruments measuring magnetic susceptibility at high temperatures the electromagnet

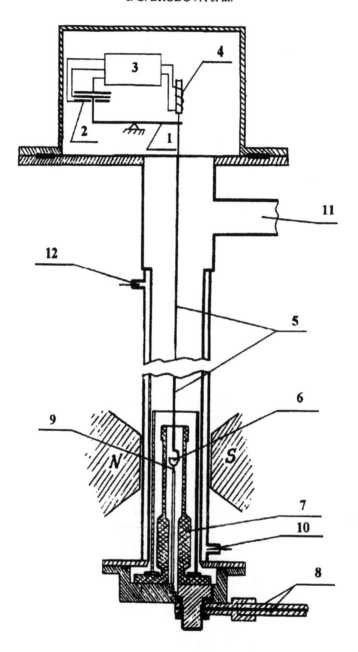

Figure 1.4 Experimental setup for measuring magnetic susceptibility of melts by the Faraday method: (1) pendulum scales, (2) three-electrode capacitor, (3) feedback system, (4) stationary solenoid, (5) suspension, (6) crucible filled with melt, (7) graphite crucible, (8) bifilar current supply, (9) tungsten–rhenium thermocouple, (10) cooling water outlet, (11) vacuum system outlet, (12) cooling water inlet.

with polepieces of special shape is typically used as a source of external magnetic field, and the force is measured with microbalances of various types. Magnetometers used by various researchers differ mainly in the construction of these main parts.

The scheme of one of the modern magnetometric devices constructed by Gol'tyakov is shown in Fig. 1.4. The unique part of this setup is the pendulum balance showing extremely high sensitivity to variations in the force acting upon the specimen. The pendulum is mechanically connected to the central plate of a three-electrode capacitor. The antiphase alternating exciting voltages, taken from the secondary winding of the low-frequency transformer, are applied to the outer plates of the capacitor. As a result of the shift of the central plate, its potential varies from zero value in the middle position of the plate to the exciting voltage in the position close to one of the outer plates. If this signal is detected synchronously with the exciting voltage, the direct current signal proportional to the shift of the central plate appears at the output. The oscillatory system of the pendulum is also connected to the core made of a ferromagnetic Sm–Co alloy and can change its position inside the stationary solenoid. The output direct current signal received by the solenoid restores the equilibrium position of the pendulum. Thus, measuring the output signal allows the magnetic field force acting upon the specimen to be determined.

This magnetometer provides accuracy of determination of absolute values of magnetic susceptibility in the temperature range from 20°C to 2200°C at a level of ± 1.3%, and the accuracy of susceptibility determinations varies on account of temperature variations at a level of about ± 0.7%.

The experimental methods listed in this section of the book are the basic methods for detection and tentative analysis of structural rearrangements in high-temperature metallic melts. They provide the minimum amount of experimental data allowing one to determine the temperature or composition at which significant changes in the melt structure occur and to answer the question of whether this change is associated with rearrangement of the short-range order or large-scale inclusions. Experiments show that the short-range order rearrangements are most pronounced in the temperature and concentration dependences of electrical resistance and magnetic susceptibility. Variations in dispersity and in the volume fraction of nanometer-sized colloid particles are clearly seen in the corresponding viscosity dependences. The sensitivity of gamma-ray absorption density measurements is high enough to identify rearrangements in the short-range order. At the same time, they permit one to reveal signs of the melt separation in the gravitational field, which is related to the presence of disperse inclusions in the melt.

In parallel with measurements of the properties listed above, useful information on structural rearrangements in melts can be obtained by studying the surface tension, speed and damping of ultrasound, saturated vapour pressure, activities of the melt components, thermoelectromotive force, Hall effect, and other characteristics.

When measuring the temperature or concentration dependence of any property, false anomalies may appear owing to the imperfection of the measurement method or instruments. Only the combined study of a series of properties by independent techniques may insure the researcher against the ensuing errors. Moreover, as indicated above, these measurements must be supplemented by direct diffraction studies of the structure of the short-range order and by the investigation of the structure and properties of crystalline or amorphous specimens obtained from the melts in different structural states.

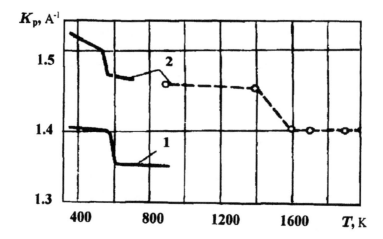

Figure 1.5   Variation with temperature of the position of the first peak of the structure factor $S(k)$ in liquid metals: (1) cesium [11], (2) rubidium (———, [12]; ━ ━ ━, [13]).

1.2   Structural Transformations in Liquid Metals

By the structure of a liquid metal is usually meant the structure of its intrinsic short-range order, i.e. the coordination in positions of atoms around a given atom, which is conditionally specified as a central one. In liquids, in contrast to solids, the nearest surrounding of an atom varies with time. Thus, we can only say of the time-averaged structure of the short-range order. Its basic characteristics are the radii R_i of the coordination spheres, i.e. the distances at which neighbouring atoms are located with maximum probabilities, and the corresponding coordination numbers z_i, i.e. the time-averaged numbers of neighbouring atoms located at these distances. The structural transformation in a liquid metal is thought to be the rearrangement of its short-range order, which is accompanied by considerable changes in R_i and z_i and occurs in a narrow temperature range. Experimental information on such processes can be derived either from direct diffraction studies or by studying the temperature dependences of structure-sensitive properties.

At the initial stage of the study of the structure and properties of liquid metals, the measurements were conducted with comparatively large temperature step and their accuracy was low. Naturally, under such conditions the researchers approximated the limited sets of experimental values by smooth analytical dependences exhibiting no anomalies and the question of the possibility of structural transformations in melts did not arise. The situation changed in the 60 and 70s, when a significant increase in the accuracy of measurements and a decrease in their laboriousness were achieved as a result of improving experimental methods. This improvement was immediately followed by a series of publications in which the signs of structural rearrangements in liquid metals were noted.

In particular, an increase in the temperature coefficient of the speed of sound in liquid indium at a temperature of about 980°C was reported in [9]. The study of the properties of liquid gallium showed a change in the temperature coefficient of electrical resistance in the

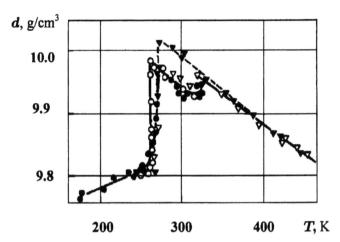

Figure 1.6   Temperature dependences of the density of liquid bismuth (after Ivakhnenko [15]); • first heating, ○ first cooling, ▽ secondary heating, ▼ secondary cooling.

range 240–270°C and an increase in the temperature coefficient of the speed of sound at 280°C. Anomalies in the temperature dependence of the speed of sound were also reported for germanium at 980°C, for lead at 900°C, for antimony at 830°C, and for bismuth at $T > 390°C$. Anomalies on the temperature dependences of viscosity for bismuth at 300°C and for tin near the solidification temperature were also attributed to the existence of structural transformations in melts.

In some cases, anomalies on the temperature dependences of the properties of liquid metals correlate with peculiarities of their diffraction patterns. Skryshevskii [1] reported that viscosities of liquid rubidium and liquid cesium undergo anomalous changes in the temperature range 520–580°C. In the same temperature range, Sharykin et al. [11] and Astapkovich et al. [12] observed a jumplike displacement of the first maximum of the structure factor $S(k)$ for these molten metals (Fig. 1.5). High-temperature data for rubidium obtained by Franze et al. [13] give evidence of another drastic change in the position of this maximum at 1230°C (Fig. 1.5). According to Gel'chinskii and Vatolin [14], this temperature corresponds to the onset of the transition of molten rubidium into the non-metallic state, which is accompanied by anomalies in the temperature dependences of electrical resistivity, magnetic susceptibility, heat capacity, and other properties.

Figure 1.6 shows the temperature dependences of density of bismuth obtained by Ivakhnenko [15] by the gamma-ray absorption method. In melting, the density of bismuth increases by 1.8% and then another discontinuous growth of density, by 0.6%, is observed, now in the liquid state, at 325°C. Dutchak [16] revealed around this temperature the anomaly in the temperature behaviour of electrical resistance of liquid bismuth and the reduction of the first coordination number z_1. Ivakhnenko treats the results of his work [15] as evidence of the two-step transition of bismuth from the unclosely-packed rhombohedral structure inherent in the crystalline state to the comparatively close-packed melt at the melting temperature with a subsequent change in the short-range order and the transition to the more closely packed structure at 325°C.

The problem of thermodynamic stability of various structural states of a liquid metal was first discussed by Ivakhnenko [15]. It was found that stable values of density d of liquid bismuth can be obtained by holding it for a long time in the temperature range from the melting point T_m to 325°C. In the above temperature range, identical and well-reproducible $d(T)$ curves are recorded in thermal cycling of the specimen. If the melt is heated above 325°C, subsequent cooling gives a $d(T)$ curve with no anomaly near this temperature and deviating at $T < 325$°C from the previously obtained dependence. However, upon solidification and the second melting of the specimen, the $d(T)$ curve contains the same features as after initial heating. The results of the work [15] show, in the author's opinion, that the high-temperature state formed in heating the melt above the point of anomaly is prone to deep supercooling with subsequent decrease in temperature and persists as a metastable state up to the solidification of the specimen. Equally well these results might be interpreted as testimony of the transition of the metal upon melting into a long-lived metastable state with intermediate density, which was inherited from the initial loose crystalline state and is destroyed irreversibly upon heating above the indicated temperature. In any case, the possibility arises of influencing the structural state of liquid bismuth prior to its solidification by changing the temperature regime of melting. It seems likely that such a rearrangement of the melt can strongly affect the process of its solidification and further the structure and properties of the crystalline specimen being formed.

Over many years the attention of researchers has been focussed on anomalies in the temperature dependences of various properties and short-range order parameters of liquid iron near 1650°C, which were discovered first in 1969 by Vertman and Samarin [17]. No less than 15 works published during the ensuing years contained reports on such anomalies. Among the most dramatic examples we can call a drastic increase in viscosity (by 6–8%) accompanied by a nearly twofold reduction in the energy of activation of the viscous flow; an abrupt decrease in magnetic susceptibility by 10%; and a 2% increase in electrical resistivity [18]. Thermal effects are detected by Fröberg and Cakici [19] near 1600°C in cooling molten iron from 1800°C. According to the diffraction studies [17, 20, 21], the structural transformation occurring in liquid iron in the temperature range from 1630 to 1670°C is accompanied by an increase in the mean interatomic distances from 2.56 to 2.62 Å and in the number of the nearest neighbours from 10.4 to 12.5. The authors classified tentatively this transformation as a transition from a bcc-like short-range order to an fcc-like structure. As a possible cause of this transition, Vatolin and Poluchin [22] indicate the additional ionization of Fe^{2+} ions to the highest valence form (+3) or the excitation of d^6 electrons and their redistribution between collective and localized states.

Along with a great body of data on the structural transformation in liquid iron, a similar amount of works are known in which no anomalies in the temperature dependences of properties and structural parameters corresponding to this transformation were revealed, although in some cases the authors used the same or even more sensitive measuring techniques. As a result, the mere fact of the existence of this rearrangement has remained questionable for a long time. Some investigators considered the previously detected anomalies as the result of experimental errors caused by such factors as oxidation of the specimen surface, the interaction between the specimen and the crucible, distortions in thermocouple readings, etc. As one of the most probable reasons for the change in the melt structure, the significant influence of impurities on the structure and properties of liquid iron was also suggested. We will analyze this effect in greater detail in the next section of

χ, e.m.u./g

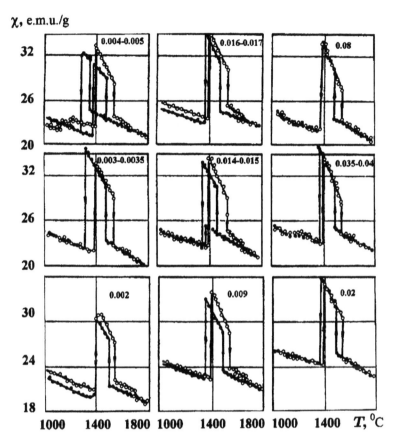

Figure 1.7   Temperature dependences of the magnetic susceptibility of iron with different oxygen content (after Sidorov *et al.* [23]): ○ heating, ● cooling; the oxygen concentration in weight per cent is indicated near each curve.

this chapter. It is only worth noting here that the change in the content of, say, oxygen in the range 0.001–0.01 at. %, which may well occur in measurements of the properties of the melt owing to its interaction with the setup atmosphere and the crucible, causes a significant change in the properties of the melt and in the short-range order parameters of liquid iron.

Only experiments performed with a thorough control of the content of chemically active impurities and, in particular, of oxygen, could give the answer to the question whether the given structural transformation is really caused by the change in the impurity concentration or it is a property inherent in pure liquid iron. The most convincing results were obtained by Sidorov *et al.* [23] who studied the temperature dependences of magnetic susceptibility of liquid iron containing various amounts of oxygen (from 0.002 to 0.08 wt. %). It was shown that the abrupt reduction in magnetic susceptibility, which was first established by Vertman and Samarin [17], is observed only in the specimens containing less than 0.02 wt. % of oxygen. As the impurity concentration decreases, this effect becomes more and more pronounced (Fig. 1.7). Consequently, the corresponding structural transformation is

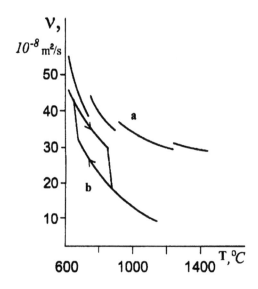

Figure 1.8 Temperature curves of viscosity of liquid aluminium: (a) [18], (b) [24].

inherent in liquid iron and the rise in the oxygen content may only suppress this transition. Ambiguous results obtained by various authors are explained by the fact that they used iron of different purity.

Measurements of the magnetic susceptibility χ were carried out by Sidorov *et al.* [23] in the range from the room temperature to 1800°C. This made it possible to obtain additional information about the influence of the structural transformation in liquid iron on the solidification process. It was established that if the melt was not heated above the temperature T_{an} of the anomalous change in χ, the $\chi(T)$ curves obtained in heating and subsequent cooling coincided and solidification into the δ phase occurred without signifi-cant supercooling. If, however, the temperature was in excess of T_{an}, the $\chi(T)$ curve obtained in subsequent cooling did not coincide at $T < T_{an}$ with the heating curve, the solidification was preceded by a deep (to 300°C) supercooling, and in some cases liquid iron solidified directly into the γ phase bypassing the δ–γ transformation "predicted" by the phase diagram (see Fig. 1.7 at the oxygen content [O] = 0.014–0.015%).

The results obtained by Sidorov, in combination with diffraction data, suggest that the metastable melt forms upon melting of iron and the short-range order in this melt retains some characteristic features of the initial δ phase. Owing to this resemblance, such a melt does not experience significant difficulties in solidification into the "parent" bcc structure. Upon heating above the point of anomalous susceptibility, the metastable state of liquid iron is destroyed irreversibly and the short-range order loses its similarity to the δ phase (becoming, possibly, γ-like, as stated in the cited works). On further cooling, the initial δ-like state of the melt is not restored and, as a consequence, the nucleation of crystals with bcc structure is hindered and starts at deep supercooling or does not occur at all. Thus, in the work [23] the possibility of changing the character of solidification of a liquid metal caused by the structural transformation in it was first established experimentally.

The features of structural transformations were also repeatedly noted in the study of the structure and properties of liquid aluminium. For example, Vertman and Samarin [17] revealed in the temperature dependence of viscosity obtained in heating the melt above the melting point three regions of abnormally rapid viscosity change: 720–770°C, 860–900°C, and 1100–1210°C (Fig. 1.8a). The temperature dependences of viscosity of liquid aluminium of various purity were obtained in a vacuum and in an inert atmosphere [24] using an alternative method of measurements: the $v(T)$ curves were first recorded in cooling the melt from the temperature 1100°C at which evacuation was made to remove the oxide surface film and then, upon solidification and remelting the specimen, the temperature dependence was recorded in heating. The results obtained in vacuum (Fig. 1.8b) point to the anomalously rapid growth of viscosity of liquid aluminium of highest purity (grade A999) in cooling in the range 880–900°C and to the inverse effect in heating in the range 1030–1080°C. The above-mentioned anomalies move toward the low-temperature region as the content of impurities increases. The study of short-range order in liquid aluminium by X-ray diffraction made by Vatolin with coworkers [20, 25] showed that in heating to 790°C all maxima of the scattering curve shift toward greater angles θ, the secondary maximum at the first peak, clearly detected at lower temperatures, disappears, but an additional maximum appears in the range of large θ. Anomalies in the temperature dependences of the short-range order parameters of liquid aluminium at higher temperatures (950–1150°C) were also noted in the X-ray diffraction study performed by Emel'yanov et al. [26].

At the same time, the study of temperature dependences of the properties of liquid aluminium in many experimental works did not reveal characteristic features which would confirm the possibility of structural transformations. In particular, the density measurements by the gamma-ray absorption method possessing the highest sensitivity to changes in the melt structure give smooth and nearly rectilinear $d(T)$ curves. In analysis of the results of other experiments in which anomalies of different kind were observed a common trend is revealed: the methods used in these experiments are quite sensitive to the presence at the specimen surface of oxide film inherent in liquid aluminium. The discrepancy between the results obtained with oxide film at the surface of the initial specimen (Fig. 1.8a) and those obtained after the removal of this film (Fig. 1.8b) shows, for example, that the change in the surface conditions may influence significantly the results of measurement of viscosity by the method of torsion oscillations of a crucible. Theoretical analysis of the influence of surface oxidation on the results of viscosity measurements by this method was performed by V.P. Beskachko. It was shown that the presence of an even very thin oxide film is equivalent to the transition to a completely different hydrodynamical problem – analysis of oscillations of a cylinder, which is not open at the top but is closed and completely filled with the liquid under investigation. When ignoring this circumstance, researchers run the risk of drawing false conclusions on the anomalous change of viscosity of the melt at temperatures corresponding to the oxide formation or disappearance, as well as to the polymorphic transformations and chemical reactions occurring in the oxide film.

Gel'chinskii [27] studied the possibility of structural transformations in liquid metals using the available data on interatomic potentials. The inference was made that even in some metals with rather simple electron shell structure (for example, in rubidium, cesium, gallium) these potentials depend in a complex way on interatomic distances. For this reason, the thermal expansion of the melt may be accompanied by fairly sharp changes in its internal energy and related properties, as well as in the structure of the nearest atomic ordering. At the same time, for such metals as aluminium, sodium, and lithium, the dependence of the interatomic potential on interatomic distance has no peculiarities. Therefore, in the author's opinion, structural transformations should not be observed in melts of these metals.

Thus, there exist metals having a tendency to structural transformations in the liquid state at definite temperatures. In some cases, the new structure of the short-range order is retained up to the onset of solidification and then the rearrangement of the melt may have a significant effect on the process of nucleation of the crystalline phase. It is not improbable that the change in the structural state of a liquid metal affects the structure and properties of crystalline specimens obtained from this metal, although we are not aware of any works in which such effects were examined. As regards liquid aluminium, the possibility of structural rearrangements in this metal seems to be questionable and requires further investigations. However, even if a conclusion is made in the course of subsequent experiments about the absence of structural transformations in liquid aluminium, it will not mean that the processing of liquid aluminium alloys with the aim to improve the quality of ingots or castings is impossible. It will simply imply that this purpose can be achieved by using another effects, in particular, by the formation of microheterogeneity in the liquid metal by introducing impurity elements, the effect that will be discussed in the next section of this Chapter.

1.3   Effect of Impurities on the Structure and Properties of Liquid Metals

Due to the insufficient purity of initial materials, the interaction with refractory materials and the smelter atmosphere, or special microalloying, liquid metals contain about 0.0001–1.0 at. % of impurities*. It is known that even microscopic amounts of impurities may influence greatly the properties of metals in the solid state. For example, as the total content of impurities in iron decreases to 0.001%, its mechanical strength falls by several orders of magnitude. In this section we present data showing that the introduction of impurity atoms is accompanied by anomalously violent changes in the properties of liquid metals due to the formation of specific microheterogeneity in the short-range order structure.

Until 1960–1970s the view prevailed in the theory of liquid metals that the isotherms of properties in a small concentration range should display linear behaviour, which reflects the additivity in the influence of impurity atoms on the properties of the solvent metal. In particular, the theory of scattering of conduction electrons by impurity ions suggested that addition of a second component is necessarily accompanied by the increase in the resistivity ρ of a liquid metal and that the resistivity increment $\Delta\rho$ is described by the expression [28]

$$\Delta\rho = k_F s x / e^2, \tag{1.13}$$

* Here, and henceforth in this section concentrations are given in atomic per cent.

in which s is the cross-section of scattering of conduction electrons by impurity atoms, x is the atomic fraction of the impurity, k_F is the Fermi momentum of the base metal, and e is the electron charge. The scattering by impurities was assumed to be independent of other scattering mechanisms, which resulted in the independence of $\Delta\rho$ from temperature (Mathiessen's rule). With constant s, the linear relationship between the additional resistivity $\Delta\rho$ and the atomic fraction x of the impurity component follows from (1.13).

The effect of impurities on the molar volume, surface tension and viscosity of melts was investigated theoretically less rigorously, but the then existing treatments of these phenomena also led to the linear dependences of properties on x. For this reason, attempts have been undertaken to classify impurity elements according to their surface activity or their influence on the self-diffusion coefficient, viscosity, and electrical resistance by comparing effects caused by the introduction of equal amounts (usually 1%) of such elements into a base metal. Moreover, even the notion of "viscous activity" of an impurity was introduced.

It was established in the course of experimental studies that formula (1.13) describes reasonably well the effect of an impurity on the electrical resistance of liquid copper and aluminium but it is less suited for other solvents [28]. Agreement between the theory and experiment was improved by taking into account the influence of volume effects on the electron scattering cross-section s; however, even with this correction the strong temperature dependence of $\Delta\rho$ was not consistent with Mathiessen's rule. Moreover, formula (1.13) is in obvious disagreement with the drop in resistance of a solvent in the case of introduction of lithium, silver, and almost all polyvalent metals in mercury; introduction of cadmium, aluminium, and gallium in tin, etc.

The linear behaviour of isotherms of density, viscosity, and surface tension in the region of dilute solutions was long beyond doubt. In accordance with theoretical predictions, the "viscous activity" of, say, 3d impurities in aluminium correlated with maximum heats of mixing. The study of the influence of these elements on viscosity ν_{Fe} of liquid iron showed that transition metals of IVa, Va and VIa periods, stabilizing the ferrite structure in the solid state, raise ν, whereas metals soluble in gamma-iron (cobalt and nickel) lower it slightly.

However, with an improvement in the sensitivity of experimental methods and by decreasing the concentration step it was found that in most cases the isotherms of properties are essentially nonlinear in the region of low impurity concentrations. Esche and Peter [29] showed that the introduction of oxygen is accompanied by a strong reduction of the surface tension σ_{Fe} of iron only at $x_0 < 0.2\%$, whereas further increase in x_0 decreases σ_{Fe} only slightly. It was found before long that the viscosity of iron is also highly sensitive to a change in the content of sulfur, carbon [30], oxygen, nitrogen [31], and aluminium [32] within tenths of per cent and responds weakly to the subsequent growth in the concentration of these elements. Anomalously strong changes in viscosity upon introduction of the first portions of potassium in mercury is reported in the monograph by Ubellohde [33]. An abrupt reduction of electrical resistance of liquid iron upon introducing carbon and oxygen additives was detected in [31] and [34], respectively.

Even more unexpected results were obtained in a series of works where extremes were revealed on the isotherms of properties of dilute metallic solutions. It seems that the dependence of this kind was first obtained in the study of viscosity of Fe–C–O melts [35], where the minimum in η at a carbon concentration of 0.7% and the maximum at 1.8% were observed. Later on, the non-monotonic initial segments of viscosity isotherms were found in the systems Fe–C [36–39], Fe–Ni [40–42], and Fe–V [43]. In the case of Fe–C, similar dependences were obtained in the measurements of density [44] and short-range order parameters [45]. It is of interest that at near-liquidus temperatures, the initial part of the density isotherm for Fe–C melts exhibited oscillatory behaviour: two minima and two maxima were clearly seen in it, with the amplitude of oscillations reducing with an increase in the carbon content.

At the initial stage of the investigation of impurity effects in liquid metals the positions and heights of extremes on the isotherms of properties obtained by various authors are widely different. Moreover, in some subsequent works [71, 72] linear concentration dependences without anomalous segments were detected once again. This did not permit the researchers to make definite conclusions about the influence of small amounts of the second component on the properties of even such well studied systems as Fe–C and Fe–O. Kushnir attributed the above discrepancies to the presence of uncontrolled impurities. To distinguish the effect of the third component on the properties of Fe–C–O melts, the regression analysis of the data obtained on a series of samples of different chemical composition was done. As a result, the conclusion was made that the change in oxygen content in the system may cause even qualitative changes in the dependence of its properties on carbon concentration. Similar results attesting to the non-additive effect of impurities were also obtained for Fe–S–P–O melts. However, samplings of specimens in [53] were inadequately representative and further experimental studies with the use of highly sensitive methods and careful control of the content of impurity elements were required to check the validity of these conclusions. A series of such experiments was undertaken at the Ural Polytechnical Institute in 1970–1980s. Below we will present a brief compendium of the results obtained with dilute metallic solutions, the impurity and matrix atoms of which have different electron shell structures.

Dilute mutual solutions of iron triad metals. The study of viscosity of Fe–Ni and Fe–Co alloys by Bodakin *et al.* [46, 47] revealed the minima on the isotherms at 2–5% and maxima at 10% of the second component. These features were attributed to variations in the structural state of the melts on alloying. The existence of such rearrangements was immediately supported by the results of the X-ray study performed by Klimenkov, who revealed anomalies in the concentration dependences of the position of the first maximum of the structure factor at 5–6% Ni in Fe and at 6% Co in Fe. The Fourier transformation of the structure factors showed that they correspond to the minimum numbers of nearest neighbours and to the kinks in the concentration dependences of the shortest interatomic spacing.

An essential step forward in the study of impurity effects in liquid metals was made with the advent of the highly sensitive gamma-absorption method of density measurements. Popel and Kosilov [48, 49] used a special dosing device that enabled them to introduce into the initial melt a series of small weighed portions of the second component without depressurizing the instrument. As a result, a variety of new features has been reliably detected in the initial segments of isotherms of density and molar volume and some regularities have been established in the influence of the impurity on the structure of the base metal.

The data obtained from measurements of density and calculations of molar volumes for Fe–Ni and Fe–Co melts are presented in Fig. 1.9. Their joint examination leads us to conclude that the oscillatory behaviour of density isotherms is a common property of dilute mutual solutions of iron triad metals. The following regularity deserves attention: the

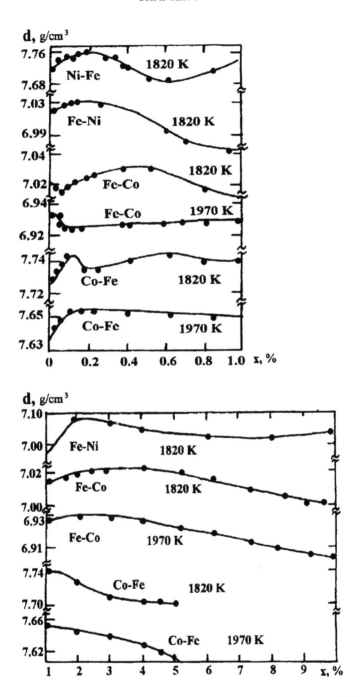

Figure 1.9   Isotherms of density of dilute mutual liquid melts of iron triad metals [48, 49].

introduction in the melt of the first portions (of order 0.01%) of the impurity with atomic number z_{im} lower than that of the matrix (z_m) is attended with the initial reduction and

28   I. G. BRODOVA *et al.*

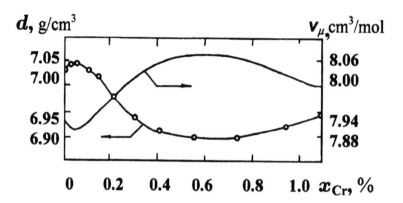

Figure 1.10   Initial portions of isotherms of density and molar volume for Fe–Cr melts at 1600°C [50].

subsequent growth of the molar volume v_μ. However, when alloying the melt with a metal with $z_{im} > z_m$, the molar volume initially increases and then begins to drop. The derivatives $|dv_\mu/dx|$ at $x \to 0$ and $x \to 1$ are proportional to the compressibilities of the corresponding solvents.

More recently, the minima of ρ and $d\rho/dT$ at 5% of the second component were revealed in the study of resistivity of Fe–Ni and Fe–Co melts. Their positions agree with the position of density extremes in Fig. 1.9. The oscillations of properties of the melts containing less than 1% of the second component are found in the isotherms of photo-conductivity of Fe–Ni and Fe–Co melts.

Chromium and scandium in iron. The effect of chromium on the properties of liquid iron was expected to be less significant than the influence of other 3d transition metal impurities due to the higher stability of its 3d electron shell. However, in the study of magnetic susceptibility of Fe–Cr alloys Singer revealed the maxima in the isotherms at 0.5% and 6% Cr separated by the minimum at 4% Cr. More recently, two series of measurements of the density of dilute solutions of this system containing up to 1% Cr were carried out [50]. In the first series, iron remelted preliminarily in a hydrogen atmosphere and containing no more than 0.02% of oxygen was used, whereas in the second the authors used iron of the same grade, containing 0.27–0.31% of oxygen, and not subjected to remelting.

The dependence of density on the chromium content x_{Cr} obtained in the first series and the isotherm of molar volume calculated from this dependence are presented in Fig. 1.10. These dependences are similar to the curves obtained for mutual dilute solutions of iron triad metals. The results also confirm the above-mentioned regularity in the succession of appearance of the first extremes of molar volume as a function of the ratio of atomic numbers of the impurity and the solvent. The position of the minimum in the isotherm of density correlates with the maximum of magnetic susceptibility obtained by Singer. In the initial portion of the isotherm (below 0.1–0.2% Cr), the results of various experiments were in poor agreement with each other. However, in all cases the introduction of the first portion of chromium was followed by a rapid decrease in the melt density. The rate of density change decreased with the growth of x_{Cr}.

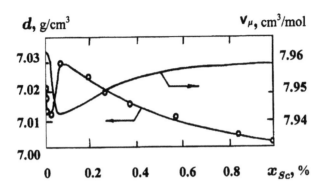

Figure 1.11   Initial portions of isotherms of density and molar volume for Fe−Sc melts at
1550°C.

The peculiarities in the change of molar volume of the melt with the components of close
atomic radii and with low chromium concentrations revealed in these works may be
attributed only to the substantial rearrangements of the atomic ordering in liquid iron near
Cr ions. The divergence of the results of the first and second series of experiments shows
that the increase in the oxygen content in the matrix affects the character of these rearrangements.

More recently anomalous changes of properties in liquid iron with a chromium content
of about 0.5% were observed in measurements of resistivity, viscosity, optical characteristics, and magnetic susceptibility χ. In the latter case, the four distinct oscillations
detected in the $\chi(x_{Cr})$ curve did not disappear when the oxygen content increased to 0.2%.
Diffraction studies showed that in iron with 0.5% Cr the concentration dependence of the
first peak of the structure factor passes through a maximum. It is of interest that maxima in
the isotherms of the self-diffusion coefficients and thermal expansion of dilute Fe−Cr
solutions were observed at this concentration in the solid state as well.

In contrast to chromium with its stable electron configuration, scandium has the most
incomplete 3d electron shell which determines its high chemical activity. It was interesting
to compare the influence of scandium on the properties of liquid iron with the effect of other
3d transition metals. With this aim the concentration dependences of density, surface
tension, viscosity, and magnetic susceptibility of Fe−Sc melts were examined at the
scandium content up to 1%.

The initial portions of the isotherms of density and molar volume at 1550°C, with
extremes located near 0.025% and 0.06% Sc, are shown in Fig. 1.11. The succession of
appearance of extremes in Fe−Sc melts differs from that in iron melts with other 3d
impurities having the atomic number less than that of the matrix. This fact may be related
to the size factor: the radius of a scandium atom is much greater than the radii of 3d metals,
including iron. Minima at 0.06% Sc are also detected in the isotherms of other properties.

More recently the measurements of viscosity of Fe−Sc melts were carried out in the
range of scandium concentrations from 0 to 3.5%. In addition to the features presented in
Fig. 1.11, another five extremes were found in the $v(x_{Sc})$ curve, with amplitudes decreasing
with the growth of the scandium content.

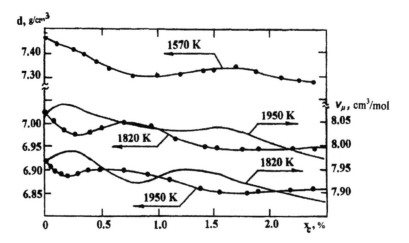

Figure 1.12   Concentration dependences of density and molar volume of Fe–C melts [55].

Carbon and boron in liquid iron. The above results indicate that the alternate succession of extremes in the initial segments of isotherms of properties is revealed after introducing into a molten 3d metal the impurity of another 3d metal, irrespective to the 3d shell structure of the latter. It was interesting to compare these effects with the influence on the properties of liquid iron of non-metallic or semimetallic additives – carbon and boron atoms – that do not have an unoccupied 3d shell.

As was mentioned above, the oscillating concentration dependences of density of liquid metallic solutions with a minimum at 0.7% of the second component and a maximum at 1.8% were first recorded in the study of Fe–C melts [44]. More recently Gel'd *et al.* [52] observed the peak in the isotherms of magnetic susceptibility of dilute liquid solutions of this system at $x_C = 0.7\%$. X-ray diffraction study of Fe–C melts revealed specific features in the concentration dependences of the position of the first maximum of the structure factor and of the height and position of the first peak of the radial atomic distribution function near a carbon content of 1.4% [53]. In [54] the maximum was found to exist at $x_C \approx 1.2\%$ in the isotherm of photoconductivity at the wavelength $\lambda = 0.6$ μm. It should be noted that in the cited works no more than four experimental points were obtained near each maximum, which impeded the location of these features. Moreover, in some cases the amplitude of oscillations was comparable with the experimental error.

More accurate concentration dependences of density of dilute Fe–C solutions were obtained by the gamma-ray absorption method [55]. The oxygen concentration in the specimens studied was lower than 0.01%. The results are presented in Fig. 1.12. A minimum at 0.25% C and a maximum in the range 0.7–0.9% C are distinctly seen in the density isotherm at 1550°C. An additional extremum at 1.4% C is observed in the isotherm of molar volume. An increase in temperature to 1680°C results in a certain smoothing of the curve in the region of extremes and in the shift of the first density minimum to 0.14%. It is interesting that similar concentration dependences of density were also obtained upon solidification of the melts for dilute solid solutions of this system, with the exception of the first density minimum which is offset toward higher carbon concentrations.

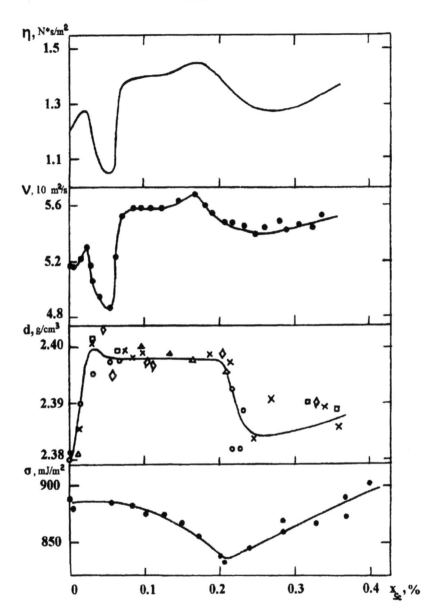

Figure 1.13   Influence of small scandium additions on dynamical viscosity η, kinematic viscosity ν, density *d*, and surface tension σ of liquid aluminium at 700°C [60].

Extremes of density and other properties of Fe–C melts in the region of low concentrations of the second component were observed in the works [56–58]. In this case two complete oscillations were detected in the concentration dependences of magnetic susceptibility and resistivity. Some discrepancies in the positions of resulting extremes may be related to the different oxygen content in the specimens.

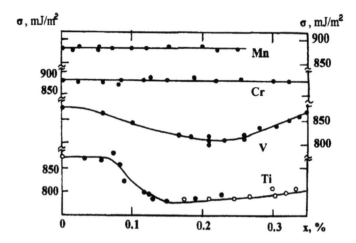

Figure 1.14   Influence of titanium, vanadium, chromium, and manganese on the surface tension of liquid aluminium at 700°C.

Extremes in the initial portions of the isotherms of density and molar volume were also revealed in the study of dilute solutions of boron in liquid iron [59]. The first extremum is found near 0.2% B.

Scandium in liquid aluminium. Analysis of the above data concerning the impurity effects in liquid 3d melts leads us to the conclusion that the oscillating concentration dependences of properties of dilute solutions are not related to peculiarities in the structure of electron shells of impurity atoms. To answer the question whether the presence of partly localized d electrons plays a fundamental role in this phenomenon, it was necessary to study the effect of impurities on the properties of a non-transition metal. In a series of succeeding works aluminium was chosen as a representative of such metals.

Popel *et al.* [60] investigated the influence of scandium (the first metal of 3d series with incompletely filled d shell) on density, viscosity, and surface tension of liquid aluminium at 700°C in the concentration region up to 0.4% Sc. The results are presented in Fig. 1.13.

We call attention to the sharp and non-monotonic variations in the measured properties with the growth of x_{Sc} which point to significant rearrangements in liquid aluminium around scandium atoms, which cover a few coordination spheres. The direct evidence for such rearrangements is the results of the subsequent X-ray diffraction study of Al–Sc melts. The authors observed a rapid change in the heights of the first and second peaks of the structure factor and an increase in the first coordination number with increasing the scandium content to 0.04–0.05%. At a scandium content of about 0.15–0.20%, the structural characteristics return to the values inherent in pure aluminium. The above features are equally pronounced at 780°C and 1050°C, which points to the thermal stability of disturbances introduced by the impurity.

Titanium, vanadium, chromium, and manganese in liquid aluminium. Surface tension, viscosity, and density of dilute Al–Ti, Al–V, Al–Cr, and Al–Mn solutions were studied to determine to what extent the disturbance introduced by a 3d impurity in liquid aluminium is related to the incomplete filling of its d shell.

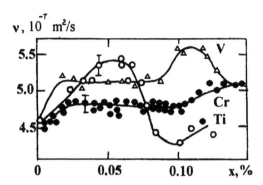

Figure 1.15   Influence of titanium, vanadium, and chromium on viscosity of liquid aluminium at 750°C.

Figure 1.14 shows the initial portions of the isotherms of the surface tension σ. Attention must be given to the similarity of the $\sigma(x)$ curves obtained in alloying Al with Ti, V, and Sc (the latter has been previously studied in [60], see Fig. 1.13), with a nearly horizontal initial segment of the curve followed by a minimum at the concentration of the second component equal to 0.20–0.25%. Isotherms of surface tension of the Al–Cr and Al–Mn systems show up in another way: the surface tension of Al–Cr melts grows insignificantly with increasing transition metal concentration, but remains virtually unchanged upon introducing Mn in Al.

The concentration dependences of viscosity of Al–Ti, Al–V, and Al–Cr melts are plotted in Fig. 1.15. When considered together with the isotherm of viscosity of the Al–Sc system (Fig. 1.13), they also point to the different influence of Sc, Ti, and V, on the one hand, and Cr, on the other, on the structure of liquid aluminium. The initial segments of isotherms of the viscosity v of Al–Sc, Al–Ti, and Al–V alloys (up to 0.25%) are non-monotonic and deviate significantly from the additive straight lines, whereas the $v(x_{Cr})$ curve attests to the much weaker disturbing effect of chromium on the matrix melt. Significant rearrangements of the system around titanium atoms are also evidenced by the existence of an extremum on the density isotherm at $x_{Ti} = 0.04\%$ (Fig. 1.16). Here, the d value differs from the aluminium density by 2.3%, i.e. it exceeds by a factor of three the response of the matrix to the introduction of the same amount of scandium.

Anomalous changes in viscosity and density of liquid aluminium reflect significant changes in the melt structure around impurity atoms. This inference was confirmed by the diffraction study of dilute Al–Ti and Al–Fe solutions, in which extremes are revealed in the concentration dependences of structural characteristics obtained at the content of transition metals of about 0.05–0.06%. Rearrangements occurring in liquid aluminium near 3d impurity atoms are related to the changes in the state of the electron subsystem, which is evidenced by the extremes in the concentration curves of the components of the complex refractive index and by the oscillations in the isotherms of light conductivity. We infer from diffraction data that the introduction of iron causes a more substantial rearrangement of the atomic ordering in the melt than the alloying with titanium and scandium.

Joint consideration of the results obtained in the study of dilute solutions of transition metals in liquid aluminium shows that the effect of a transition metal on the structure of

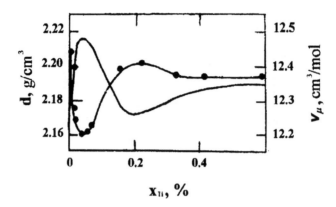

Figure 1.16   Influence of titanium on the density and molar volume of liquid aluminium at 900°C.

liquid aluminium depends non-monotonically on the number N_d of d electrons. As the d shell is filled, this influence decreases from Ti to Cr and Mn and then begins to grow again. Scandium takes a special position in this series; having $N_d = 1$, it is less active than Ti (a similar feature was noted above for solutions of 3d metals in liquid iron). The above regularity correlates with the dependence on N_d of the Hall constant R_H for solid solutions of group IV transition metals in aluminium: according to [61], at impurity concentrations of about $10^{-3}\%$ and at temperatures up to 50 K the function $R_H(N_d)$ has a parabolic shape with a minimum at $N_d = 5$. The only exception is scandium, the dilute solutions of which have lower Hall constants than the solutions of titanium in aluminium. The similar influence of 3d impurities on the structure and properties of aluminium in the liquid and solid state made it possible to apply some inferences obtained in the solid state physics to the theoretical analysis of this phenomenon in melts.

Cadmium in aluminium. The results of the study of properties of aluminium–transition metal alloys showed that anomalies (extremes or oscillations) are observed in the isotherms of structural characteristics and properties in the region of low content of the second component even in the case when the solvent is a simple metal. It was interesting to compare the effect of transition metals on the structure of liquid aluminium with the effect of an impurity element with no vacancies in the d electron shell. Popel *et al.* [62] chose cadmium possessing a simple $5s^2$ electron configuration with completely filled inner shells as such an element.

They studied the density and surface tension of Al–Cd melts containing up to 0.3% Cd at a temperature of 700°C. The results are presented in Fig. 1.17. Attention must be given to the significant negative deviations of molar volumes from additive values; they reach a maximum at 0.1% Cd and are not inherent in the solutions with the dominant interaction of like atoms, to which the given system belongs. The isotherm of surface tension also has a maximum near 0.1% Cd; its deviation from the dependence specific to perfect solutions reaches 4% at this composition. These data indicate that a significant change in the structural state of the melt takes place as the above composition is approached.

To summarize the discussion of the experimental results obtained in the study of properties and short-range order of dilute metallic solutions, we should emphasize that

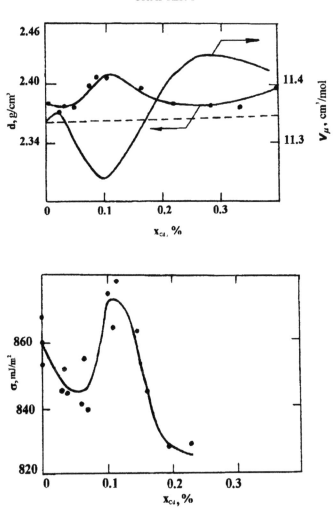

Figure 1.17   Isotherms of density d, molar volume v_μ, and surface tension σ of diluted Al–Cd solutions at 700°C [62].

extremes or oscillations in the isotherms of these properties were observed in all systems under investigation independently of the structure of electron shells of atoms of impurity and matrix. Thus, in the theoretical analysis of phenomena related to the introduction of impurities in liquid metals, we should prefer concepts that are applicable equally well to melts of any nature.

1.4   Cluster Nature of Impurity Effects in Melts

We now turn to the analysis of the data presented above. The results obtained by Belashchenko [32] in the study of electrodiffusion in metallic melts are of great importance for an understanding of the causes of the existence of anomalies in the isotherms of properties of dilute solutions. It was found that the effective charge of cadmium dissolved in sodium ranges from −80 to −6 as the cadmium content varies from 0.001% to 1%. This is equivalent to a rapid increase in the scattering cross-section s_{Cd} as x_{Cd} falls below 0.1%. Based on these data, the author of the work [32] concluded that the form of existence of the impurity varies with changing the cadmium content in the above limits. It was assumed that impurity atoms are capable of forming stable complexes with matrix atoms, which raise electrical resistance and viscosity of the solution. The origin of these complexes was attributed to the high mutual affinity of the components (inherent in compound-forming systems), or to the ionization of dissolved impurity atoms. The possible existence of the impurity in the melt in the form of ions with different charges predetermines, according to [32], the "polygeneity" of solutions, i.e. the possibility of exhibiting properties differing in the type of concentration dependences in various compositional and temperature ranges. When analyzing the effect of impurities on the kinetic properties of metals, Belashchenko suggested that the change in the structural state of the solvent near the impurity ion should also be taken into consideration. This idea was developed in more recent studies.

To estimate the influence of complex formation on viscosity of diluted solutions, the authors of the works [63, 64] used Einstein's formula for the suspension of rigid spheres:

$$\eta = \eta_0(1 + 2.5\Phi), \tag{1.14}$$

where η and η_0 are the dynamical viscosities of the suspension and the dispersive medium, respectively, and Φ is the volume fraction of disperse particles. It was assumed in this case that disperse particles are the complexes consisting of the impurity atom and z atoms of the solvent (z is the coordination number). Comparison of the calculated and experimental η values showed that formula (1.14) describes well the viscosity of Sn−Ni melts at the nickel concentration up to 2% and at relatively low overheating above the liquidus.

Assuming that the size of impurity complexes R_c may exceed the radius of the first coordination sphere and that the electron scattering cross-section is $s = \pi R_c^2$ (as in the case of spheres opaque for electrons), Belashchenko estimated the R_c values from the data on electrical resistance of a series of systems with strong interaction of unlike components and obtained $R_c \approx 0.1 - 0.2$ nm. These R_c values are too small (less than the radius of the first coordination sphere) to represent the real dimensions of the complexes. To agree the R_c estimates with the adopted model, the author assumed that complexes are partially transparent for electrons.

Mott [65] came to the problem of the existence of extremes in the isotherms of electronic properties of impurity-containing melts from different positions. He pointed out that if the solvent has a deep minimum in the curve of density of electronic states $N(\varepsilon)$, such as that in the curve of mercury, impurities can affect the electrical resistance by changing the form of the $N(\varepsilon)$ curve or position of the Fermi level ε_F. For example, the introduction of In, Sn, Bi, and other elements in mercury is accompanied by a decrease in the width of the forbidden zone in the solid phase and the depth of the minimum in the $N(\varepsilon)$ curve in the liquid phase, which is not compensated by a decrease in the free path length. According to [65], two processes may occur on dissolving alkaline metals in mercury: (1) the formation of "impurity atom−solvent" complexes, which is attended with the decompression of mercury and the increase in the depth of the minimum in the $N(\varepsilon)$ curve; (2) the transition of valence electrons into the conduction zone of mercury, which results in the increase in ε_F. At low impurity concentrations, the first effect dominates and electrical resistance ρ of the melt grows. Further addition of an alkaline metal shifts ε_F upward on the energy scale and ρ begins to drop.

The mechanisms suggested by Mott cannot pretend to explain most of the anomalous properties of liquid metals with impurities presented above because they are applicable only to melts with specific energy dependence of the density of electronic states. The interpretation of experimental data based on the similarity of the initial portions of isotherms of properties of liquid and solid solutions proved to be much more universal. We have already called attention to this similarity when discussing the results obtained in the study of the Fe–Cr and Fe–C systems. Moreover, there is a variety of works in which oscillations were observed in the initial segments of isotherms of various properties of dilute solid solutions. Among the earliest publications concerning the effects dependent on the introduction of 0.01-0.1% amounts of impurity in transition metals, we can mention variations in the self-diffusion coefficients [51], thermal expansion, Young's modulus [66], optical constants, resistivity, and parameters of the fine structure of X-ray spectra [67]. More recently similar dependences were also detected by other authors. In particular, the data on neutron elastic scattering in Fe–V, Fe–Cr, Fe–Mn [68], and Pd–Fe solid solutions [69] give evidence of intense and non-monotonic variations of the magnetic moment on impurity atoms at an impurity concentration of about 1%. Near the same composition, Ishino and Muto [70] observed specific features in the isotherms of resistivity of V–Ta and V–Fe alloys.

Gurov and Borovskii suggested the following model to qualitatively explain the results of their studies. A single impurity atom possessing an excess electric charge distorts the energy spectrum of electrons of the matrix (primarily the zone of valence electrons and the unoccupied d zone of a transition metal) in the local region with a radius of the order of $10q$, where q is the screening length equal to 0.05-0.1 nm. This causes the spatial redistribution of collectivized s and d electrons near the impurity atom. Moreover, the inner energy levels of matrix atoms are also shifted slightly, which causes an additional interaction of these levels with the impurity (the induced impurity effect). As a result, a stable "ordered block" forms in the region of the size indicated above. As the concentration of impurity atoms increases, neighbouring "blocks" begin to overlap. This causes the redistribution of electrons at the boundaries of these blocks and, therefore, a change in the effective charge of impurity atoms and neighbouring atoms of the matrix. The solution structure becomes more uniform. The authors of [51, 66] relate extremes on the isotherms of properties to the onset of interaction between the blocks. This enables the radii R_b of the ordered blocks to be estimated from the experimentally determined extremum positions:

$$x_{ex} = 3a^3/(4\pi N_{cel} R_b). \qquad (1.15)$$

Here a is the lattice parameter and N_{cel} is the number of atoms in the unit cell. By applying (1.15) to the isotherm of molar volume of Fe–C solid solutions (an fcc lattice, $Z = 4$, $a = 0.367$ nm) shown in Fig. 1.13, we obtain that its maximum at 0.8% C corresponds to $R_b = 0.72$ nm and the minimum at 1.7% C, to $R_b = 0.56$ nm.

Assuming the similarity of electronic properties of the alloys in the solid and liquid states,

Popel *et al.* [60] suggested that disturbed regions with the structure of short-range order differing from the structure of the order in metal–solvent are also formed in melts. In the following these regions will be referred to as clusters. The enhanced strength of inter-atomic bonds inside the clusters results in their motion as single objects in the volume of the liquid.

Let us analyze in this context the results of the study of viscosity and density of dilute Al–Sc solutions as described above. Following A. Ubbelohde [33], we can relate the anomalous viscosity variations to the obstruction of the viscous flow in a certain part (Φ) of the melt volume caused by the cluster formation. The particular structure of the cluster is not essential; it is only important that all its atoms move consistently in the melt.

To the first approximation, the clusters may be imagined as rigid spherical particles of radius R_c. Taking into account only the hydrodynamical interaction of particles with each other and with the suspending liquid, Batchelor [71] derived the expression for the effective viscosity η of particles' suspension, which refines Einstein's formula (1.14):

$$\eta = \eta_0(1 + 2.5\Phi + 6.2\Phi^2). \tag{1.16}$$

Here η_0 is the viscosity of the dispersion medium and Φ is the volume fraction of suspended particles. In the case under consideration it is natural to assume that η_0 is equal to the viscosity η_{Al} of pure aluminium. Then the volume fraction of clusters can be estimated from the difference between η_{Al} and viscosity of the alloy. The results of calculating the dynamical viscosity η of Al–Sc melts with the use of kinematic viscosity v and density d obtained in [60] are presented in Fig. 1.13. The initial portion of the η isotherm is characterized by drastic viscosity changes. In the range $0.02\% < x_{Sc} < 0.05\%$ the viscosity of alloys is lower than η_{Al}. It is possible that in this concentration range viscosity depends strongly on factors of a non-hydrodynamical nature, which have not been taken into account in the derivation of formula (1.16). Thus, only data in the region $x_{Sc} > 0.05\%$ can be interpreted within the framework of the model considered.

The results of the calculation of the volume fraction Φ of clusters are presented in Fig. 1.18. We see that Φ also varies with x_{Sc} non-monotonically, remaining within the limits 2–4%. The quantity Φ is determined by the expression $\Phi = (4/3)\pi R_c^3 n_c$, where n_c is the number of clusters in the unit volume and R_c is the radius of clusters. At low scandium contents, when impurity ions do not interact virtually with each other, equivalent groupings arise near each of the ions. In this case n_c coincides with the number density n_{Sc} of scandium atoms in the melt and we obtain the following expression for the cluster radius:

$$R_c = (3\Phi/(4\pi n_{Sc})^{1/3}. \tag{1.17}$$

The results of the calculation of the concentration dependence of the cluster radius are also presented in Fig. 1.18. They show that microgroupings of radius ~0.6 nm exist in the melt at a scandium concentration of 0.06%. Even if their density does not differ from the aluminium density, they should contain about 50 atoms. With further growth of x_{Sc}, the radius of suspended particles reduces to 0.35 nm and, starting from $x_{Sc} = 0.2\%$, remains on that level.

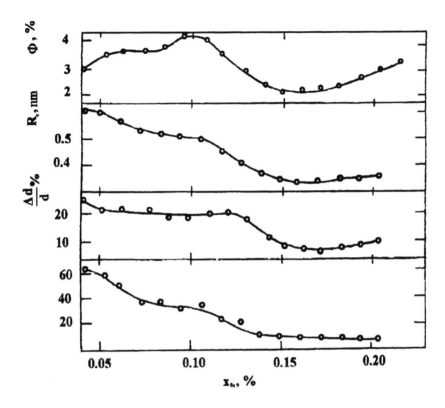

Figure 1.18 Variation of cluster parameters with scandium concentration in liquid aluminium at 700°C.

Naturally, the model of rigid suspended particles is very far from the real situation: most researchers consider the cluster as a liquid-like formation having no interphase boundary with the surrounding melt. However, the particular shape of the cluster is not too important for the hydrodynamical analysis of the system. Hushin [72] estimated the upper and lower bounds for viscosity coefficients for the isotropic emulsion of liquid particles of arbitrary shape with internal viscosity η_{in} differing from the viscosity η_0 of the suspending liquid and with zero interphase tension at the interface. For $\eta_{in}/\eta_0 > 4$, the volume fraction of such objects calculated from the effective viscosity of their suspension differs only slightly from the Φ values obtained for the suspension of rigid particles. For this reason, the results of calculations of Φ and R_c presented in Fig. 1.18 are not devoid of the physical sense.

If the change in the density of liquid aluminium caused by scandium additive is ascribed to the formation of clusters, the volume fraction and size of which are determined by expressions (1.16) and (1.17), the difference Δd between the densities of the alloy and pure solvent can be represented as $\Delta d = \Phi \Delta d_c$, where Δd_c is the difference between the cluster density d_c and the density of aluminium d_{Al}. This enables one to estimate the cluster density at various scandium concentrations and to calculate, from the known density and radius of such a grouping, the number N_c of aluminium atoms in it (Fig. 1.18). At $0.08\% < x_{Sc} < 0.18\%$, the cluster density exceeds d_{Al} by about 20%, while at $x_{Sc} > 0.18\%$ it falls and from then on the ratio $\Delta d_c/d_{Al}$ remains close to 8%. The number of atoms in the cluster decreases monotonically with increasing x_{Sc} from 64 at 0.06% Sc to 9–10 at $x_{Sc} > 0.2\%$.

Since the vacancy concentration in liquid aluminium does not exceeds 1–2%, the resulting Δd_c values cannot be explained only by the absence of vacancies in clusters. Evidently, the interatomic spacings in such a grouping are significantly reduced (by about 7–8% at $0.06\% < x_{Sc} < 0.18\%$ and by 2–3% at $x_{Sc} > 0.2\%$). The non-monotonic behaviour of isotherms of density and viscosity in the segment $x_{Sc} < 0.06\%$ which terminates in the formation of clusters in the melt, attests to the complicated multistep nature of this process.

Thus, the results of measurements of the properties of Al–Sc melts confirm the existence of clusters with a size close to that of the ordered blocks in solid solutions. This allows us, following [51, 66], to relate extremes in the isotherms of properties with different stages of interaction between clusters. Relationship (1.17) complies with the conditions under which clusters fill the entire volume of the specimen. Let us convert (1.17) to a form that does not contain the characteristics of the crystal lattice and therefore is suitable for the description of melts:

$$x_{ex(cp)} = 3v_\mu/(4\pi N_a R_c^3) \tag{1.18}$$

(v_μ is the molar volume and N_a is Avogadro number). When condition (1.18) is satisfied, the close packing of clusters is attained with the maximum possible space filling factor $\xi_{cp} = 0.64$. According to the percolation theory, anomalies in properties depending on the degree of electron localization may also be observed in the region $\xi_p < \xi_{cp}$ corresponding to the beginning of electron transfer between clusters in contact. In the three-dimensional lattice of uniform spheres percolation begins at $\xi_p = 0.3$. Therefore, extremes on the isotherms can also be revealed at the concentrations $x_{ex(cp)} = \xi_p x_{ex(dp)}/\xi_{cp}$. The results of the estimate of the radius of closely packed clusters and the percolation threshold from the experimentally determined positions of extremes of molar volume of the systems under investigation are given in Table 1.1. This table also contains the numbers of particles in the cluster, which were estimated on the assumption of close atomic volumes in the interior of the cluster and outside it. These estimates show that clusters with 10–20, 30–50, 100–200, 400–600, and 900–1000 atoms are present in the systems under study but clusters with intermediate numbers of particles are nearly absent. This is indicative of a definite periodicity in the filling of coordination spheres.

The model proposed by Gurov and Borovskii is qualitative in its nature. The ordered block, or cluster, manifests itself as a single object having no interior structure. Of prime importance for the more rigorous study of the microheterogeneous distribution of impurities in metals is the concept of virtual bound states of electrons in impurity atoms put forward by Friedel [73]. According to [73], the bottom E_0 of the potential well related to the impurity atom may be located in the region of positive energies of the continuum spectrum. At definite E_0 values, the so-called virtual bound state (VBS) with a finite lifetime appears. Superimposing on the continuum states, the VBS resonates with them, which results in the expansion of the level E_0 into a band of width ΔE localized near the impurity atom. This virtual level is responsible for the strong scattering of electrons with energies close to E_0. It may be assumed that electrons are trapped at the VBS levels for a some time $\Delta t \sim h/\Delta E$ (h is Planck's constant). The VBS level changes the electron density near the impurity atom, which oscillates with the spatial period

$$\lambda = h/(2m_eE_0)^{1/2}, \tag{1.19}$$

where E_0 is the VBS energy with reference to the bottom of the conduction band and m_e is the effective electron mass. The distances R at which these oscillations are damped can be estimated from the expression

Table 1.1 Cluster size and the number of atoms in clusters estimated from the extremes in the isotherms of molar volume.

System, temperature		Cluster radius, nm		Number of atoms	
		close packing	percolation threshold	close packing	percolation threshold
Fe–Co	0.05	1.85	1.24	2000	599
	0.5	0.86	0.57	200	60
	1.0	0.68	0.46	100	30
1550°C	3.0	0.47	0.32	33	10
Co–Fe	0.1	1.45	0.97	999	299
	0.2	1.15	0.77	499	149
	0.7	0.76	0.51	143	43
1550°C	3.0	0.47	0.31	33	10
Fe–Ni	0.03	2.19	1.47	3332	998
	0.17	1.23	0.82	588	176
	1.0	0.68	0.46	100	30
1550°C	2.0	0.54	0.36	50	15
	8.0	0.34	0.23	12	4
Ni–Fe	0.15	1.26	0.84	666	199
1550°C	0.63	0.78	0.52	159	48
Fe–Cr	0.06	1.73	1.16	1663	498
1660°C	0.6	0.81	0.54	166	50
Fe–Sc	0.025	2.33	1.56	3999	1196
1550°C	0.06	1.74	1.16	1664	499
Fe–C	0.25	1.08	0.72	400	120
1550°C	0.8	0.73	0.49	125	37
	1.4	0.61	0.41	71	21
Al–Sc	0.03	2.47	1.65	3331	996
700°C	0.21	1.29	0.86	476	142
Al–Sd	0.03	2.47	1.65	3331	996
700°C	0.1	1.65	1.10	999	299
	0.3	1.20	0.77	333	100

$$R \approx h/(2m_e \Delta E). \qquad (1.20)$$

Therefore, with the given parameters of the virtual bound states E_0 and ΔE, we can determine the scale of perturbations of electron density near the impurity and the internal periodicity of this perturbation.

The results of calculations of the width and position of VBSs in aluminium crystals with impurities [74–77], and the experimental data obtained from spectroscopic measurements [78] are presented in Table 1.2. This table also contains the spatial periods λ and the lengths R of the corresponding perturbations of electron density. Attention should be given to the fact that the R values are close to the radii of clusters in melts estimated in the

Table 1.2 The positions E_0 and the widths ΔE of the virtual bound states in impurity atoms in solid aluminium and the results of the estimate of the spatial periods λ and the lengths R of the corresponding electron density perturbations.

Impurity	E_0, eV	ΔE, eV	Reference	λ, nm	R, nm
Ti	12.1	4.3	[75]	0.35	0.59
	12.6	5.7	[76]	0.35	0.51
	11.8	3.26	[74]	0.36	0.68
V	11.5	2.9	[75]	0.36	0.72
	11.6	3.4	[76]	0.36	0.67
	11.8	–	[77]	0.36	–
	11.4	2.56	[74]	0.36	0.77
Cr	11.2	2.1	[75]	0.37	0.85
	11.3	2.9	[76]	0.36	0.72
	11.0	2.05	[74]	0.37	0.86
	11.0	1.8	[75]	0.37	0.91
Mn	11.0	2.3	[76]	0.37	0.81
	11.4	–	[77]	0.36	–
	10.9	1.4	[78]	0.37	1.04
	10.6	1.63	[74]	0.38	0.96
	10.8	1.5	[75]	0.37	1.00
Fe	10.8	2.0	[76]	0.37	0.87
	10.3	–	[77]	0.38	–
	10.7	1.5	[78]	0.37	1.00
	9.8	1.25	[74]	0.39	1.10
Co	10.4	1.2	[75]	0.38	1.12
	10.5	1.5	[76]	0.38	1.00
	10.1	1.6	[78]	0.39	0.97
	8.2	0.80	[74]	0.43	1.37
Ni	9.8	0.92	[75]	0.39	1.28
	9.8	0.6	[76]	0.39	1.58
	9.2	1.2	[78]	0.39	1.12

hydrodynamical approximation and determined from the positions of extremes in the isotherms of molar volume (Table 1.1). The intersection of the virtual level with the Fermi level in aluminium occurs in passing from vanadium to chromium, or between chromium and manganese, if we take into account the multiple scattering of electrons [79]. It is possible that this peculiarity determines the absence of some anomalies in the isotherms of properties of dilute Al–Cr and Al–Mn melts (Figs. 1.14 and 1.15). Comparison of R and λ values indicates that impurity atoms produce in their neighbourhood multilayered perturbations of electronic density. As a consequence, the related clusters have a complex structure consisting of several sparsely spaced, but comparatively closely packed, atomic layers. Such a structure of clusters results in the multistep pattern of interaction between them at small mean distances between impurities, as well as in the increase of the number of specific features on the isotherms of properties.

Unfortunately, the works cited in Table 1.2 do not contain the VBS parameters for scandium impurity in aluminium. However, Lautenschlager and Mrosan [80] calculated for 3d impurities in aluminium the electron charges localized inside and outside the effective zone of action of the model impurity potential. It was shown that the scandium is more effectively screened by conduction electrons than the titanium atom and therefore exerts a lesser perturbation on the structure of the aluminium matrix. This is consistent with the data presented above, which show that scandium has a weaker effect on the density and viscosity of liquid aluminium than titanium.

The parameters of virtual bound states on impurities in transition-metal crystals are calculated with lesser accuracy than those for non-transition metals [81, 82]. They agree in the order of magnitude with the data in Table 1.2. In all the cases considered above the λ and R values calculated from these parameters are close to each other, which is evidence of the multilayered nature of clusters. The characteristic scale of perturbations of electron density lies more frequently within the limits 1–2 nm. Therefore, the perturbation covers a few coordination spheres, i.e. the region containing 100–1000 atoms.

The existence of such comparatively large-scale perturbations around impurity atoms in crystals is confirmed by the results of many experiments. In particular, anomalously high values of the Hall constant and magnetoresistance of solid solutions of aluminium with 10^{-3}–$10^{-2}\%$ Ti and V point to the strong localization of conduction electrons at the impurity sites [83]. As the temperature of these alloys drops, significant anisotropy in the electron scattering appears, which is explained in [84] by the formation of an ultradisperse phase comprised of impurity clusters. The measurements of hyperfine fields in the Fe–P system showed that the perturbation of their characteristics by the phosphorus atom extends approximately over five interatomic distances. The temperature dependences of parameters of the magnetic component of diffuse neutron scattering in dilute solutions of Si, Ge, V, Ti, Ni, and Co in iron are consistent with the calculations based on the molecular field approximation only if the effect of the impurity on 8 coordination spheres of neighbouring iron atoms is taken into account.

The most convincing evidence of microheterogeneity of dilute solid solutions are obtained from nuclear magnetic resonance (NMR) studies. The introduction of an impurity atom in a pure metal is accompanied by the formation of an electric field gradient at the surrounding lattice sites, which is sufficient for the loss of resonance by a certain number N of matrix nuclei. The number N is determined by the decrease in the intensity of NMR lines. It was established that the number of non-resonant neighbours is 38 for the Ga impurity in Cu [87]; 98 and 167 for Zn [88] and Cu [89] in Al; 63, 75, 126, and 194 for Cr [90], Mn [91], N and O [90] in V, respectively. If impurity atoms have at least a small magnetic moment, N sharply increases (for example, $N = 850$ for Mn in Al [89]). The indicated values are calculated from experimental data obtained at low temperatures. As temperature increases, the disturbing effect of an impurity on simple metals is significantly lowered. However, even near the melting point the number of non-resonant neighbours is equal to some tens [88]. In transition metals the electric field gradients at impurity atoms may not only decrease but even increase with increasing temperature [70]. It is likely, therefore, that a strongly pronounced impurity microheterogeneity can also be observed in high-temperature melts.

The effect of impurities on the electronic and atomic structure of liquid metals is theoretically investigated less intensively than the influence of impurities on metallic crystals. The possibility of the formation of magnetic clusters in melts of 3d-transition metals with impurities is substantiated in the works by Dolgopol with coworkers [92]. Fairlie and Greenwood [93] modified the model of rigid spheres as applied to an impurity ion dissolved in a simple liquid metal. The structure of the liquid was described by the Percus–Yevick structure factor and the parameters of the local environment of impurity atoms were determined by minimization of the free energy of the system. Distinct oscillations of the electron density around the Cl^- ion in metallic sodium are detected at distances up to 0.8–1.0 nm. The authors of [93] calculated the heats of dissolving chlorine, fluorine, and oxygen in liquid sodium and obtained values reasonably consistent with experimental values. Singer and Sandratskii with collaborators [94] suggested a way of estimating the radius of an impurity cluster based on the Friedel sum rule. They considered an impurity atom dissolved in the "jelly" of free electrons modelling the matrix melt. By varying the damping factor R for oscillations of the impurity potential, the authors solved the Schrödinger equation and determined the phases of scattered waves. They used for the cluster radius R_c the value of R for which the difference in Friedel's sums, corresponding to the impurity and matrix potentials, was equal to the difference in their valences. The resulting R_c values for Mo, W, Nb, and Ta impurities in pure iron are 0.46, 0.57, 0.58, and 0.59 nm, respectively – agreeing reasonably with our estimates presented in Table 1.1.

The most reliable source of experimental information on the scale of impurity microheterogeneity of metallic melts is the results of optical measurements. Shvarev suggested that the position and the width of the virtually bound states should be determined from the photon energy dependences of high-frequency conductivity for the matrix and the alloy. Using these data, the radii of perturbations of electron density around Ni and Fe impurity atoms in Al (1.2 and 1.4 nm), Ni in Cu (1.5 nm), and Ge in Fe (1.1 nm) were determined. The resulting values are in good agreement with the estimates of cluster radii from the positions of extremes of optical characteristics. Two virtually bound states corresponding to the s–d and d–d resonances and to the clusters with radii 1.2 and 0.5 nm were revealed. In Shvarev's opinion, the existence of clusters of different types in the melt is the main cause of the appearance of several extremes in the isotherms of properties. An alternative point of view was argued by Sidorov who relates oscillations in the isotherms to the consecutive overlapping of oscillations of the impurity potentials with increasing content of the second component. This concept is consistent with the shape of isotherms of molar volume of 3d-transition metal based alloys: the first extremes corresponding to large distances between impurity atoms have a smaller amplitude in comparison with subsequent extremes: they seem to be related to the overlapping of distant low-amplitude oscillations of potentials.

Thus, non-monotonic changes in the properties of dilute metallic solutions exist both in the liquid and in the solid state. The theoretical study of crystals with impurities points to the formation of disturbed regions with the complex multilayered structure around foreign atoms. The experimental results presented in this section support the existence of such formations (clusters) in melts, the sizes of which are approximately identical in liquid and solid solutions. This makes it possible to use, on the one hand, the results of calculations of impurity perturbations in crystals for the estimate of the scale of microheterogeneity of melts, and, on the other, to determine the content of the components providing the optimal level of service characteristics of alloys upon solidification using the results of the study of properties of liquid metals with impurities. The experimental results illustrating the efficiency of the last line of investigation for solving the problems of practical metallurgy are described in Section 3.4.1.

1.5   The Structure of Concentrated Liquid Metallic Solutions

Liquid alloys are commonly classified according to the appearance of their phase diagrams. The following main types of binary and multicomponent systems are distinguished:
(1) those with a continuous series of solid solutions;
(2) systems with intermetallic compounds;
(3) eutectic systems;
(4) systems with a miscibility gap in the liquid state.

As with pure metals, diffraction experiments and analysis of physico-chemical properties are used for the study of the structure of liquid alloys. The first data on the scattering of X-rays pointed to the similarity of diffraction patterns of molten and crystalline specimens and led researchers to the idea that along with the standard short-range order a certain ordering also exists in the arrangement of atoms of different sorts. The degree of this ordering may be different – from slight deviations from the random distribution in the systems with a continuous series of solid solutions, to the formation of compounds or regions consisting chiefly of atoms of one sort in the systems of types 2 and 3, 4, respectively. The interpretation of compositional dependences of properties of crystalline alloys developed by Kurnakov [95] and based on the estimation of the appearance of phase diagrams proved to be quite useful also for liquid metallic solutions.

The simplest distribution of atoms of different kinds is realized in *melts with a continuous series of solid solutions* (type 1). The curves of scattering of X-rays, electrons, and neutrons by such objects are interpreted with reasonable accuracy as a result of diffraction from the regions with random distribution of the component atoms [1]. The isotherms of their physico-chemical properties have been assumed previously to be smooth monotonic curves corresponding to the small negative or, more rarely, positive deviations from perfect solutions.

However, a closer examination of concentration dependences of properties of alloys of this type revealed features, which are not consistent with the idea that these alloys are close to perfect solutions. For example, Roll and Motz [96, 97] carried out a study of electrical

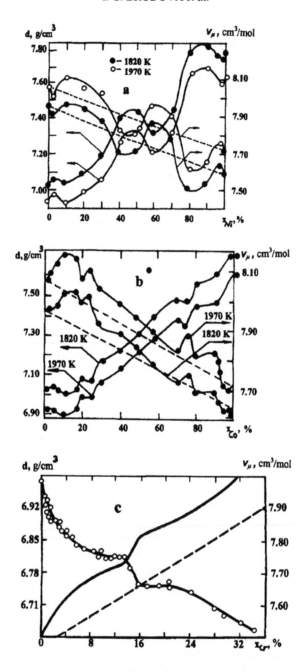

Figure 1.19 Isotherms of density and molar volume of (a) Fe–Ni [48], (b) Fe–Co [49], and (c) Fe–Cr [50] melts. Dashed lines correspond to the additive concentration dependences of molar volume.

resistance of the Ag–Au and Au–Cu systems and obtained isotherms with several extremes which are fairly symmetrically located on the concentration axis and are clustered near the concentration ratios 1:2 and 2:1. More recently, complex concentration curves were obtained in the study of viscosity of Fe–Cr melts and surface tension of Fe–Ni and Fe–Co melts. These data point with certainty to the existence of some rearrangements of the structure on alloying.

Quite representative results were obtained in the study of density of the Fe–Nu, Fe–Co, and Fe–Cr systems by the gamma-ray absorption technique [48–50]. They show that significant structural changes may occur in melts of this type with changing composition, in spite of their similarity to perfect solutions. The isotherms of density and molar volume are shown in Fig. 1.19. Attention must be given to the extremes of molar volume at 10–15, 45, 60, and 80% Ni in Fe; at 12–14, 20, 25, 70, 77, 80, and 90% Co in Fe, as well as to the anomalously rapid increase in the molar volume with increasing the Cr concentration in Fe from 13 to 16%. The deviations of the molar volume from additive values and the amplitude of the anomalies listed above do not exceed 2%. This confirms the similarity of these solutions to perfect solutions. However, the observed deviations considerably exceed experimental errors and, therefore, reflect changes in the melt structure on alloying. Note that extremes on the isotherms of molar volume for the Fe–Ni system and especially for the Fe–Co system are distributed on the concentration axis in a nearly symmetric manner relative to the equiatomic composition and bear a certain resemblance to isotherms of electrical resistance of systems with a continuous series of solid solutions [96, 97]. The amplitude of extremes of volume in the temperature range of interest is virtually independent of temperature.

More recently additional experimental data were obtained, which support the existence of anomalies on the isotherms of properties of melts of the type considered here. For example, the oscillating dependences of excess molar volume on composition with anomalies near the mole fractions 0.2, 0.3, ... of the second component, similar to those shown in Fig. 1.19, were detected by Okajima and Sakao in the study of Bi–Ag and Bi–Cd melts [98]. Bazin et al. [99] observed the maximum of electrical resistance in the Fe–Co system near 10% Co, while in [100, 101] the anomalous changes in resistance and magnetic susceptibility were found at 10–15% Cr in liquid iron. Kudryavtseva et al. [102] showed that in the latter case the extremum in ρ is related to the abrupt reduction in the number of collectivized electrons per atom from 1.0 to 0.9, i.e. with perturbation of the density $N(\varepsilon_F)$ of electron states at the Fermi level. The review [103] gives evidence for the universal character of the non-monotonic behaviour of the isotherms of properties for type 1 melts.

It is of interest that anomalies of properties of such melts correlate, as a rule, with peculiarities in the concentration dependences of properties of solid solutions, which points to the common character of structural changes occurring on alloying in the liquid and solid phases. For example, the thermoelectric power of Fe–Ni alloys experiences perturbations at 50 and 80% Ni [104], the mean magnetic moment on the Fe atom passes through a maximum at a chromium content of about 15% [105], while in the region between 10 and 15% Cr the peak in the Mössbauer spectrum of Fe–Cr alloys corresponding to Fe atoms with no Cr atoms in the nearest neighbourhood disappears [106]. A close investigation of the concentration dependences of properties of the solid and liquid solutions with a concentration step of no more than 1% was made by Andronov et al. [107]. In all the systems studied they obtained oscillating isotherms of properties. In some cases oscillations obey rather rigorous periodicity with extremes observed at compositions corresponding to the integer ratios of the component concentrations. As in the experiments performed in [48, 49], the amplitude of oscillations of properties remained virtually unchanged with increasing temperature in the ranges under investigation.

Andronov *et al.* [107] explain these results by the existence in liquid and solid solutions of concentration inhomogeneities with different proportions of unlike atoms and by the change in the types of these inhomogeneities at concentrations corresponding to the maximum probability of forming microgroups of a particular type. It is highly probable that the reason for such rearrangements is the perturbations of density of electron states at the Fermi level, near compositions at which the numbers of like atoms in the first coordination spheres of atoms of any component change by unity (this effect is theoretically explained by Kim [108]). The authors of [107] estimate the mean number of atoms N in such groups (which play the role of "molecules" of the solution-forming type) from the period of change of properties with composition. The application of this approach to the isotherms of molar volume plotted in Fig. 1.19 gives $N = 3$ for Fe–Ni melts and $N = 8$–9 for Fe–Co and Fe–Cr melts, i.e. the scale of the chemical ordering in these melts does not surpass the radius of the first coordination sphere. The presence of microinhomogeneities of such kind in Fe–Ni and Fe–Cr solid solutions at high temperatures is confirmed by the results of magnetic measurements [105] and diffuse electron scattering experiments [109].

It may be assumed that the above-mentioned atomic microgroups are formed in the process of alloy melting, when the alloy is still in the liquid state, but persist upon solidification, thus determining to a great extent the properties of the ingot. If this hypothesis is correct, the compositions corresponding to the extremal values of properties of cast metal can be derived from the positions of extremes in the isotherms of properties of melts. However, as far as we know, no attempts of such kind have been undertaken, and it is still impossible to estimate the outlook for this approach for the above purposes.

Melts of *systems with intermetallic compounds* (type 2) are likely to be classified as "physically opaque" liquids with developed microheterogeneity. Diffraction patterns of the specimens corresponding or close in composition to intermetallic compounds usually differ qualitatively from analogous curves of pure components [1], which points to the different structures of their short-range order. The distinctly pronounced anomalies – maxima of electrical resistance, viscosity and activation energy of viscous stream [17, 31], and minima of magnetic susceptibility, molar volume, and entropies of melting and mixing [110] – are usually observed in the concentration dependences of properties of type 2 melts at temperatures close to the liquidus. The positions of extremes coincide, as a rule, with compositions corresponding to the compounds with highest strength known in the solid state. These extremes are flattened, however, as temperature rises. In the study of electrodiffusion, a change in the effective ion charge is seen at the same compositions [28]. If the liquidus line of the system has no acute peaks, a fact which points to the low thermal stability of compounds, then inflections (but not extremes) or maximum deviations from additive dependences are observed in the isotherms near concentrations corresponding to such peaks. However, in many cases the positions of such features do not coincide with the compositions of compounds existing in the solid state [110].

The experimental results listed above are consistent with Kurnakov's conception of the retention in liquid melts of partially non-dissociated compounds and of the possibility of qualitative assessment of the degree of their dissociation from the profiles of concentration and temperature dependences of properties, and from the liquidus curvature near the composition corresponding to the given compound [95]. Esin obtained an explicit expression for the radius of the curvature of the liquidus line as a function of the dissociation of the compound in its melting. Using the dependence of the equation of this line on the interchange energy ω, which characterizes in the model of regular solutions the energetically non-equivalent character of interactions of like and unlike atoms, Regel and Glasov derived the formula relating the degree of dissociation α to ω and the melting temperature T_m and obtained for semiconductor compounds the values of α at T_m in the range from 0.1 to 0.45 [111] and for solutions of intermetallic compounds between 0.1 and 0.7 [20].

Whilst a full consensus on the possibility of retaining compounds in the liquid phase, the form of their existence in the melt is not quite clear. In particular, in the works of Gel'd *et al.* [112], the molten compounds were considered to consist of stable quasimolecules (for example, FeSi). By contrast, Danilov believed that their structure "should not be characterized by the formation of molecules, which would cause the intensity maxima to emerge at angles much lower than those in pure metals, but by the appearance of ordered regions, in which an alternate sequence of the component atoms takes place similar to the sequence of these atoms in crystals of the alloy" [113]. Ubbelohde [33] treated compounds in the melt as "associative complexes" of fluctuation origin. The same viewpoint on the cause of the formation of In_2Bi and $InBi$ groups was held by Styles [114] who estimated their lifetime as $\sim 10^{-7}$ s. When estimating the size of associates, the researchers obtained values no less than 2 nm, although the possibility of the fluctuation origin of a great number of such large-scale formations in melts overheated by tens of degrees above T_m is doubtful. Samarin and Vertman [17] assumed that fragments of compounds in the melt are "permanently existing groupings", which retain the hereditary short-range order of solid compounds and are destroyed only at high overheating above the melting point. Similar views were reasoned in [115, 116], where additional attention was given to the possibility of retaining the homogeneous conditions attained after such an overheating of the melt in the course of subsequent cooling down to the solidification point. It was also conjectured in these works that the degree of microheterogeneity of the melt affects the value of its supercooling prior to solidification.

The temperature dependences of the density of nickel and iron borides were measured in [59, 117] by the gamma-ray absorption method in the range from room temperature to 1900°C. A sharp reduction in the density of the molten Fe_2B compound at 1650°C was attributed to the destruction of complexes of stoichiometric composition and to the transition of the system into the state of a regular solution. The $d(T)$ curves obtained by the authors of these works in heating and subsequent cooling coincide, which gives evidence for the reversibility of the above rearrangement.

Thus, it is impossible to retain the more uniform condition of the specimen, attained upon its significant overheating above the melting point, by lowering the temperature of molten intermetallic compounds with moderate rates down to the onset of solidification. This makes problematical the possibility of governing the structure of solidified compounds by changing the temperature regime of their melting in standard smelters. It is not inconceivable, however, that the high-temperature states of liquid intermetallics can be inherited by solidified samples upon cooling with high rates typical of the processes of powder metallurgy or melt spinning, thereby modifying the structures and properties of solid samples.

The *systems of eutectic type* (type 3) have attracted the attention of researchers for more than one century in connection with the specific features of phase transformations at eutectic points, on the one hand, and due to the great practical significance of these alloys, on the other. Even today much remains to be seen in the phenomenon of crystallization of a homogeneous solution at a strictly determined temperature T_e and composition x_e into the two-phase crystal structure with developed interface. From the 30–40s onwards, predisposition of eutectic melts to such phase transitions was often attributed to their microsegregation, i.e, to the lack of a complete mixing of the components on the atomic level [118].

The concept of microsegregation is based on the results of the X-ray diffraction study of liquid Sn–Zn, Sn–Pb, and Bi–Pb eutectics performed by Danilov and Radchenko [119]. They revealed that the first maximum of the scattering curve $I(s)$ for these melts at temperatures close to T_e is branched into two nearly equal side maxima, which points to the presence of two different structures in the liquid. Following Word, who took similar diffraction patterns from organic eutectic systems, the authors of [119] suggested that there is no perfect mixing of the components in liquid eutectics near the melting point and the melt consists of atomic groups typical of the liquid α and β phases, i.e. it has a quasi-crystalline structure. To verify this assumption, they calculated the $I(s)$ functions for such a system by summing up the intensity curves specific for pure components and obtained good agreement with the experimental dependence. More recently similar results were obtained for other systems using the methods of X-ray [20, 120–123], electron [124], and neutron diffraction [125, 126]. It was pointed out that the features of microsegregation are inherent not only in melts of eutectic composition, but also in melts of other concentrations of the components. In systems with complex eutectics, one or both types of groups forming the microsegregated melt may have the structure of the corresponding molten compounds [127, 128]. At the same time, in some studies of the short-range order of eutectic melts, scattering curves corresponding to the random distribution of unlike atoms have been obtained [129–132].

As far back as the 1930s, Kurnakov examined the "composition–property" dependences for a number of eutectic-type liquid solutions and concluded that these curves have no specific features at eutectic concentrations or around them [95]. However, anomalies in the isotherms of properties were often observed in metallic melts near x_e. For example, Klyachko and Kunin [133] revealed the minimum of surface tension σ for Pb–Sn and Pb–Bi melts at the eutectic composition, which was retained on heating above T_e up to 100°C. In the same temperature region, the positive $d\sigma/dT$ values are revealed for alloys close to the eutectic composition. Roll *et al.* [134–138] observed anomalies in the isotherms of resistivity ρ and $d\sigma/dT$ near the eutectic concentration for a variety of melts. Geguzin and Pines [139, 140] reported the existence of a maximum of the energy of mixing for Pb–Sn melts and a minimum value of the heat capacity jump in melting Bi–Sn and Bi–Pb alloys at x_e, while Evseev and Voronin [141] revealed the minimum of the entropy of fusion for Sn–Pb alloys at the eutectic composition. There are also data on the existence of minima of viscosity [110] and density [142] near x_e. These data were usually interpreted as evidence of the microsegregation of melts which increases as the eutectic composition is approached.

Gaibullaev and Regel [143] carried out an exhaustive study of the temperature dependences of electrical resistance in systems with a simple and a complex eutectic and found anomalies in the $\rho(T)$ curves on overheating above T_e by 100–400°C, which were interpreted as a result of the transition of alloys from the microlaminated to homogeneous state. The data on the sharp rise of the coefficient of diffusion of Si in liquid silumin at $T \approx T_e + 500°C$ [144] are consistent with this treatment.

However, at least the same number of experimental works exist in which no anomalies were found in the isotherms of properties near the eutectic composition. We can mention the studies of molar volume v_μ [145, 146], surface tension [147, 148], viscosity, magnetic susceptibility, and electrical resistivity [149]. At the same time, in some cases peculiarities are noted at compositions corresponding to the limiting solubility of the components in the solid state (for example, the minima of ρ [149] and the maxima of v_μ [150]). It was established that the microsegregation effects are most prominent not at the eutectic concentration, but at compositions close to the equiatomic one [120, 141, 151], which are characterized by a kink or a horizontal segment in the liquidus line.

To clear up the possible causes of microlamination of eutectic melts, it was important to estimate correctly the sizes of regions enriched with various components. The most convincing results were obtained by centrifuging such systems. As far back as 1946, Bunin reported the carbon enrichment in the zone of cast iron adjacent to the centrifuge axis, which is indicative of the formation of associations containing no less than 1000 atoms of this element [152]. Later on these experiments were repeated by Vertman with coworkers [153], who measured the radial gradient of carbon concentration in samples solidified by melt spinning and estimated from it the radius of carbon groups at about 10 nm, i.e. much greater than the characteristic scale of the short-range ordering. Shortly thereafter Kumar [154–156] obtained new evidence for the large-scale inhomogeneity of Pb–Cd and Al–Zn melts. In the specimens subjected to centrifugal separation, he observed the appearance of a radial concentration gradient corresponding to the regions of pure components of 1 to 5 nm in size, as well as a macroscopic separation into two solutions with effective concentrations differing by 10–15%. The latter effect was detected in the compositional range corresponding to the flat portion of the liquidus line. Evidence of the existence in liquid silumin of Si-rich regions with size decreasing from 4.5 nm at 700°C to zero at 1100°C was obtained by Izmailov and Vertman [157] as a result of similar experiments. These authors also hypothesized that the effective influence on the silumin structure can be achieved upon solidification by changing the melt temperature to control the range of dimensions of Si precipitates. The fact that the characteristic inhomogeneity scale for liquid eutectics does not exceed 10 nm is also evidenced by the data obtained in the experiments on light scattering in organic systems of this type [158] and by the results of the study of small-angle scattering of X-rays in metallic melts [125].

Whilst the mere fact of large-scale microheterogeneity of eutectic melts is beyond question, the causes of its existence have not been clear for long. The authors of [159, 160] treated microlaminations of like atoms as steadily existing associations, which arise due to the predominant interaction of such atoms. Indeed, most of the eutectic systems formed by metallic components have positive heats of mixing ΔH [141, 151]. This points to the stronger interaction of like atoms compared to unlike atoms. However, the data systematized in [118] show that in these cases the ΔH value is too small to justify such a lamination. Moreover, most of the eutectic melts for which convincing evidence of microheterogeneity was obtained have $\Delta H < 0$. Therefore, they have no energetic stimuli to be separated into regions enriched with different components. Grigorovich [161] assumed that microlaminations arise as a result of fluctuations, thus representing dynamical groups. However, it is readily established that such large-scale compositional fluctuations are possible only in the very narrow region near the eutectic point: as the distance from it increases, their probability rapidly declines, whereas signs of microsegregation are observed in a far more extensive range of states.

The colloidal pattern of the structure of eutectic melts, which was first formulated by Klyachko and developed by Vertman with coworkers [162] and most consistently by Zalkin [163], deserves special consideration. These authors considered eutectic melts as classical colloidal systems with particles of size about 1–10 nm. It was suggested [162] that such a state is implemented only in the limiting case of solutions with strong covalent interaction of particles of one of the components (for example, Fe–C), whereas in other eutectics the microheterogeneity has a fluctuating nature. Vertman *et al.* [162] treated the colloidal state of liquid cast iron as non-equilibrium, but argued at the same time that this state can be restored, though in the form of more disperse particles, in cooling of the Fe–C solution homogenized by overheating above the liquidus line. Zalkin considered any eutectic melts to be lyophilic two-phase systems (microemulsions) with low interphase tension at the boundaries of fine particles. According to his ideas, the transition from such a state to the homogeneous state on heating is reversible: the initial microheterogeneity is reestablished upon cooling the homogenized melt. The fact that microsegregation effects are most pronounced near the equiatomic composition is consistent with the emulsion model. Microemulsion in the region above the liquidus was considered in [163] to be thermodynamically stable, which immediately provoked objections in connection with the violation of the phase rule in the eutectic point. Another serious objection against the colloidal model was that the decomposition of the system into different phases in cooling the homogeneous solution must be accompanied by a decrease in entropy and an increase in the interphase free energy. In systems with small positive and especially with negative heat of mixing these effects cannot be compensated by a decrease in the internal energy.

In the last few years data have been obtained in the small-angle X-ray and neutron scattering experiments, which indicate with certainty that microlaminations of like atoms in eutectic melts do not arise and disappear by means of fluctuations, but exist as long-lived and possibly even stationary formations. For example, Bi–Cu melts were studied in [164] by the cold neutron small-angle scattering technique. It was found that the coherent scattering cross-section changes by about 40% as temperature rises from 800 to 1100°C. This is equivalent to a 7–8% reduction in the size of microscopic regions enriched with one of the components, which is a very small change for inhomogeneities of fluctuation origin. Bellisent-Funel *et al.* [165] measured the small-angle neutron scattering from Ag–Ge melts. The systematic discrepancies between the experimental and theoretical coherent scattering cross-sections revealed in this work have led the authors to the conclusion that this system cannot be described in terms of simple density and compositional fluctuations and therefore some quasistatic structures should exist in it. Huijben *et al.* [166] used the methods of small-angle scattering of X-rays and neutrons for the investigation of the structure of Na–Cs melts and came to the conclusion that long-lived compositional inhomogeneities, slightly varying in size with temperature, should exist in the liquid phase.

The causes of the formation and long life of microscopic regions rich in like atoms in liquid eutectics became clear only with the advent of the theory of metastable microheterogeneity of melts (see Section 1.7). Wide potentialities of controlling the structure and properties of eutectic alloys in the crystalline state by varying the degree of microheterogeneity of the initial melt were also established at that time.

In *systems with an immiscibility dome* (type 4), as well as in eutectic systems, indications of the separation of microlaminations of like atoms are observed in the region of single-phase states. Diffraction curves of such melts can be represented as a result of the superposition of diffraction patterns from pure components [1]. Various anomalies (in particular, viscosity and resistivity maxima [31]) are detected near the critical concentration in the isotherms of properties at temperatures exceeding only slightly the critical lamination temperature T_c. The formation of the above microlaminations in such systems is associated with the preferred interaction of like atoms. Near the critical point these laminations may arise by means of fluctuations. However, the scale of fluctuations should rapidly decrease as we recede from the critical point. Indeed, the maxima in the isotherms of electrical resistance disappear when the temperature exceeds T_c by only 2–3 K. At the same time, viscosity anomalies are observed when the alloy is overheated by tens or even hundreds of degrees above the critical point [110]. The opinion was stated that these anomalies are caused by microscopic liquid flows with a scale of ~10 nm related to the development of fluctuations in the near-critical region. However, such flows should die out together with fluctuations. Nikitin [167] considered the liquid solutions in the region above T_c to be ultradisperse emulsions. From these positions, he managed to explain maxima in the viscosity isotherms. At the same time, the possibility of retaining such micro-heterogeneous structure away from the critical point was rejected, although the measurements of small-angle X-ray scattering showed that even in low-temperature organic liquids of the type considered here, microheterogeneities of about 10 nm exist upon overheating above T_c by tens of degrees [1].

The theory of the metastable microheterogeneity of liquid metallic solutions, which appeared at a later time, made it possible to explain the large-scale microheterogeneity of type 4 melts from the same positions as the microlaminations of liquid eutectics (see Section 1.7). Based on this theory, the methods affecting the structure of these systems have been developed; they influence significantly the tendency of melts toward macroscopic lamination with decreasing temperature, as well as the structure of the ingot being formed in this process.

1.6 Transmittance of Hereditary Traits in the "Charge–Melt–Ingot" Chain and Non-equilibrium States of Liquid Alloys

Ample experimental material accumulated in metallurgical practice indicates that the structure and properties of alloys with the given chemical composition (within the limits indicated by controllable impurities) depend to a large extent on the type of charge materials used and on the way they are introduced into the system. In particular, the effect of the volume fraction of carbides in the initial cast iron on the properties of castings, the influence of grain size in cast iron, and of charge metal scrap on the structure and properties of open-hearth steel, and the effect of electrode wire on the properties of welded joints, etc. have all been established. In recent years, a set of such phenomena connected with the transmittance of structural features of the charge material via the liquid state to the ingot or casting was collectively named a "structural metallurgical heredity" [167]. In the metallurgy of ferrous metals the heredity phenomenon has been studied since 1920s. In the field of aluminium alloy production interest in this phenomenon arouse only in the 1970s after the works of Krushenko with coworkers who showed that the structure of castings can be favourably influenced by the heat and mechanical treatment of charge materials and by the introduction of premelted or finely divided alloying components.

On the other hand, it was known that manifestation of hereditary characteristics of a charge material in the ingot depends on the technological conditions of melting, primarily on the heat–temperature regime of the process, as well as on mechanical, electromagnetic and other actions on the melts. In particular, it was possible to decrease or even eliminate the hereditary effect of charge materials by multiple remelting, an increase in the melting temperature and the exposure time in the mixer, evacuation, or out-of-furnace treatment, etc.

These data showed that some structural formations inherited from charge are retained in the liquid phase and that their condition can be changed, to the point of full failure, under the influence of thermal and other actions. It was natural to expect that the presence of such hereditary "genes" and the change in the form of their existence would be revealed in the structure-sensitive properties of melts. Since the 1960s systematical studies of the influence of the structure of charge materials, methods and technological regimes of melting on the properties of melts and ingots produced from these melts have been conducted at the Ural Technical and Pedagogical universities [168, 169]. It was shown, in particular, that the viscosity and surface tension of grade 30KhGSA liquid steel obtained from direct-reduction charge are higher (and the electrical resistance is lower) than these properties for the specimens of the same steel obtained from usual metallurgical charge. These discrepancies cannot be accounted for by the different content of impurity elements, gases and non-metallic inclusions. The properties of specimens obtained from various charge materials approached each other as the melt temperature increased, which pointed to the destruction of hereditary structural constituents. An interesting result was obtained in the study of molten specimens prepared from different structural zones of a continuous ingot of steel 20. At nearly identical compositions, metal from the zone of columnar crystals had lower viscosity and lower conductivity than metal taken from the central and crust zones. It was also established that the viscosity of liquid steel of the given grade depends on the production mode and grows regularly in the line: open arc melting → vacuum arc remelting → vacuum arc remelting in the alternate electromagnetic field → electroslag remelting. Plasticity characteristics, elasticity moduli, and thermal conductivities of metal after solidification rise in the same sequence. Baum *et al.* [168] assigned the observed change in properties to the enhancement of a complex of external factors acting upon the melt in passing from open arc melting to electroslag melting.

The results obtained have led the authors of [168] to the idea that at the same temperature melts can exist in different structural states, depending on their past history. This assumption is confirmed by the branching (hysteresis) of temperature dependences of properties obtained in heating and subsequent cooling of the specimens. Belashchenko studied the properties of Cd–Sb alloys and found that viscosity was higher on cooling from 450–500°C than on the preceding heating [170]. Kononenko observed the branching of temperature dependences of viscosity and electrical resistance of Ga–(20–30%)In alloys. More recently the hysteresis of temperature dependences of various properties has been observed by Gotgil'f and Lyubimov [171, 172], Baum *et al.* [168], and others. In some cases the branching of the "property–temperature" curves occurred on cooling the metal from any temperature and their divergence increased with the increasing maximum

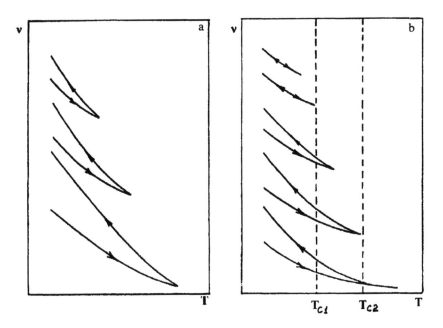

Figure 1.20   Main types of branching the temperature curves of viscosity of liquid metallic melts which were recorded on heating and subsequent cooling of the melts.

temperature of specimen heating (Fig. 1.20a). For most melts hysteresis appeared only upon heating to the definite "critical" temperature T_{c1} and increased up to a certain temperature T_{c2}; further increase in the maximum heating did not change the shape of the cooling curve (Fig. 1.20b).

The branching of temperature dependences of properties was attributed initially to irreversible changes in the content of non-metallic inclusions, gases, and other impurities [7, 173]. However, long-term careful studies of variations of these factors on heating above the liquidus line and their influence on the properties of melts and ingots have led the researchers to the conclusion that these factors play a secondary role in heredity and hysteresis phenomena. For example, Kushnir showed that wide variations in the content of such impurities as sulfur, phosphorus, and oxygen may only slightly displace the temperatures T_{c1} and T_{c2} in alloys exhibiting branching of temperature dependences. In contrast, in liquid metals showing no tendency to the hysteresis of properties branching does not arise even if the impurity content increases significantly. An understanding has gradually formed that the branching of temperature dependences is due to irreversible changes in the melt structure rather than attendant factors. For example, Belashchenko [170] explained the rise in viscosity of Cd–Sb melts by the change in the structure of short-range order from that characteristic of the initial CdSb compound to the structure corresponding to the arrangement of atoms in the metastable Cd_3Sb_2 phase formed in the solidification process.

The ambiguity of states of melts at the same compositions and temperatures showed that many of these melts are non-equilibrium or metastable. The possibility of a long-term non-equilibrium structural state of liquid metals and alloys is corroborated by the results of experiments in which time variations of their properties were recorded under isothermal conditions. Belashchenko [170] observed long-term drifts in the viscosity and electrical resistance of Bi–Tl alloys. After rapid heating (for 3–5 min.) of the specimen with 72% Tl to 650°C, its viscosity ν dropped steadily with time τ approaching asymptotically a stable value. The approximation of the ν(τ) dependence by the exponential function enabled the author of [170] to determine the characteristic relaxation time which proved to be close to 40 min. Later on time variations in viscosity of this alloy were measured in [171] at the isothermal holding of the specimen rapidly heated from the solid state and rapidly cooled from 700°C to the same temperatures. In both cases the viscosity values were stabilized in 1–3 hours and coincided with each other, which pointed to the equilibrium final state. In the study of electrical resistance of Ni–Si alloys at 1400–1550°C its instability was observed over a few hours. The above relaxation times exceed by many orders of magnitude the characteristic times describing atomic motions in the liquid state and hence the results of the experiments listed above immediately attracted the attention of researchers.

There were several ways of explaining irreversible changes and the long-term non-equilibrium of properties of melts. The belief was stated, in particular, that structural transformations in the liquid phase may follow the pattern of polymorphic transitions and that the metastable states may exist for a long time at either side of such transitions (the phenomenon examined by us in Section 1.2 [171]). This version remains disputable until now, because indications of abrupt structural changes are noted only for a limited number of liquid metals. With regard to the hysteresis phenomenon, it is observed in melts of different nature and is inherent in binary and multicomponent systems rather than in pure metals. Attempts were undertaken to explain hysteresis by the irreversible destruction in the liquid phase of the strongest directional bonds inherited from the initial crystalline specimen [28]. Such an explanation seems to be acceptable for systems retaining a considerable fraction of covalent interaction after melting (such as liquid silicon and germanium, or alloys made of these elements). However, the existence of non-equilibrium effects in alloys based on thallium, gallium, nickel, and other elements, which are "good" metals in the liquid state, calls for a more universal approach to their interpretation. In some works these effects were treated as testimony of the decreased microheterogeneity of melts, however, without specifying the last term. Chipman [174] expressed his opinion more definitely by relating the hysteresis of structure-sensitive properties of binary and multicomponent systems to the extremely slow development of processes leading to the uniform distribution of atoms of different kinds. He reasoned that even intensive mixing of a metal does not provide a complete compositional leveling in its microvolumes. This point of view was at variance with the traditional concept about the diffusion–convective mode of mixing components in metallic melts and was long ignored by most researchers.

The work by Popel *et al.* [175] provided a substantial step toward understanding the reasons for the long-term non-equilibrium of properties of metallic melts. Using the possibilities offered by gamma-ray densitometry, the authors observed time variations in the density of the irradiated zone of Fe–Ni and Fe–Co specimens of equiatomic composition, in the course of formation of liquid alloys from separate ingots of their components under isothermal conditions. It was found that the stationary value of density of a Fe–Ni melt was established at 1550°C in about 4 hours, whereas in a Fe–Co melt it was attained under the same conditions in only 30 min (Fig. 1.21). The results obtained led

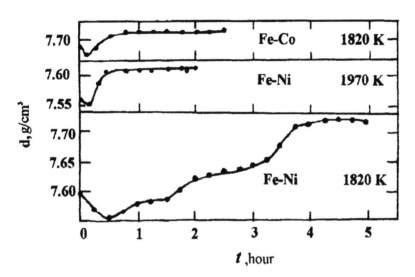

Figure 1.21   Variation with time of the density of the irradiated zone in Fe–Ni and Fe–Co alloys in the course of isothermal mixing of the components [175].

the authors to the conclusion that in this case the relaxation time is determined by the rate of mutual diffusion of the components. If the mass transfer were accomplished by atomic diffusion, the duration of the relaxation process would be much longer. (The characteristic time of diffusion to distances close to the radius of the crucible used in experiments is equal to 40 hours at 1550°C.) Moreover, since the diffusion coefficients for nickel and cobalt in liquid iron differ by no more than 10%, the relaxation times in Fe–Ni and Fe–Co alloys should be close to each other. In real experiments, however, their difference attained an order of magnitude. Certainly, the mass transfer in massive molten specimens is accelerated by convection. The convective mixing rate under identical conditions is determined chiefly by the density and viscosity of the components and by the thermal coefficients of these quantities. These parameters in Fe–Ni and Fe–Co melts are so close to each other that the derived differences in relaxation times cannot be explained by the different rates of convective mass transfer.

In an attempt to resolve the above contradiction, Popel put forward a hypothesis, according to which the factor determining the mass transfer rate at the initial stages of the process of mixing of dissimilar melts, is the macroscopic mixing of the components [175]. At this stage the system consists of two phases separated by a boundary with low interphase tension. Spontaneous dispergation of these phases may occur under such conditions, starting with the loss of hydrodynamical stability of the boundary between the phases. The distinction between the melts under study and classical emulsions lies in the fact that on account of the unlimited mutual solubility of the components, a continuous mass transfer takes place in this case across the interphase boundary, which reduces the interphase tension.

The dynamical interphase tension σ_{A-B} determines the predisposition of the system to dispergation. Spontaneous dispergation at room temperature begins only at $\sigma_{A-B} \sim 0.1$ mJ/m^2 or less. At elevated temperatures, the critical σ_{A-B} values prove to be much higher.

Experimental studies of the dynamical interphase tension in systems with unlimited mutual solubility were not carried out. For this reason, Popel *et al.* [175] used for comparative estimation of this quantity Antonov's rule, according to which $\sigma_{A-B} = \sigma_A - \sigma_B$ (σ_A and σ_B are the surface tensions of pure components at the boundary with the gaseous phase). Using the reference data for the quantities σ_{Fe}, σ_{Ni}, and σ_{Co} they obtained $\sigma_{Fe-Ni} \approx 120$ mJ/m^2 and $\sigma_{Fe-Co} \approx 30$ mJ/m^2. In accordance with this, the loss of stability of the boundary due to the reduction of σ_{A-B} in the Fe–Ni system should occur at a much later time than in Fe–Co.

The study of systems with more complicated phase diagrams confirmed the validity of this hypothesis and led the investigators to the idea of non-equilibrium or metastable colloidal states of melts, which will be discussed in the next sections of this chapter.

The study of the causes of long-term non-equilibrium states of melts, common regularities, and specific manifestations of this non-equilibrium in different systems has a practical significance for developing the optimal regimes of alloy production.

Up to now, the temperatures of heating various alloys in the process of melting are determined basically from the data on the position of the liquidus line, taking into account casting conditions, stability of refractories, and economical considerations. They rarely exceed the liquidus temperature by more than 100–200°C. In searching for resources for improving metal quality, the investigators studied the effect of overheating above the liquidus on the structure and properties of ingots and articles [173, 176]. In steel melting, the overheating had, as a rule, a positive effect, while in the aluminium alloys production it was usually accompanied by coarsening of the structure, with the exception of some cases in which, by contrast, it was possible to obtain a more favourable modified structure [142] due to the rise in the melting temperature. The detailed metallographic examination of Al–Cu and Al–Mg alloys heated to various temperatures in the molten state, which was performed by Novikov and Zolotorevskii [177], showed that the processes of dendritic segregation occur in them in different ways. Following Spasskii, they related this effect to irreversible changes in the structure of liquid metal. In the cited works, the overheating values were determined in an arbitrary way, without allowance for information on the structure and properties of melts at various temperatures.

Starting in the mid-1960s, Baum *et. al.* developed a different approach to the determination of optimal time–temperature regimes of melting, based on the analysis of temperature dependences of properties of melts. Knowing that heating a liquid metal to the "critical" temperatures (T_{c1} and T_{c2} in Fig. 1.20b) causes accelerated transition into the equilibrium state, and having established that this transition is favourable in most cases for the structure and properties of ingots and articles, these authors suggested that melts should be overheated under industrial conditions either to T_{c1} or to T_{c2} [168]. Additional studies of relaxation processes at these temperatures determine the time required for the corresponding isothermal holdings. The time–temperature regime of melting developed on the basis of experimental data has come to be known as the heat–time treatment (HTT) of a melt. The introduction of HTT in the technology of producing various steels and iron-based and nickel-based alloys showed the high efficiency of such an approach [168]. The influence of HTT of the melt on the structure and properties of aluminium alloys will be discussed in Chapter 2.

The lack of a clear understanding of the nature of hereditary formations in melts, the causes of their long-term existence at temperatures close to the liquidus temperature, and regularities of their destruction in heating to the "critical" temperature, for a long time did

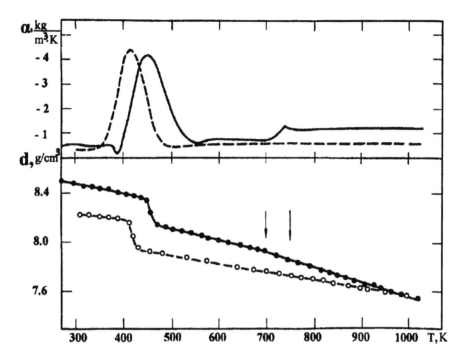

Figure 1.22 Temperature dependences of density d and thermal expansion coefficient α of Sn–26.1 at. % Pb melt obtained in cooling from 700°C (▬ ▬ ▬) and in heating after solidification of the specimen and its repeated fusion (▬▬▬) [178]. Arrows indicate the initial and final points of the anomalous portion of the heating curve.

not permit the researchers to analyze the HTT influence on the properties of melts and ingots from single positions. This, in turn, hindered the general assessment of experimental data and the transition from empirical determination of melting regimes for each particular steel grade to prediction of the efficiency of such a treatment and to *a priori* development of recommendations for various groups of alloys. With the advent in the 1980s of the theory of non-equilibrium and metastable colloidal states of melts it became possible to solve a significant part of the above-mentioned problems.

1.7   Metastable Microheterogeneity of Liquid Metallic Solutions

The phenomena of microlamination of eutectic melts, the hysteresis of temperature dependences of properties, and the structural metallurgical heredity considered above in this chapter were successfully explained from single positions after the concept of metastable microheterogeneity of liquid metallic solutions was formulated by Popel *et al.* [169] in the early 1980s.

The prerequisite for its appearance was the study of the temperature dependence of density of a Sn–Pb melt of eutectic composition [178]. The authors of this work mixed the components in the furnace of a gamma-ray densitometer at 700°C, i.e. at a temperature that significantly exceeds the eutectic point T_e and at which no evidence of microheterogeneity has previously been observed. After relaxation of the system the temperature dependence of density $d(T)$ in cooling the specimen was recorded (Fig. 1.22). In spite of the high sensitivity of the method, the authors revealed no anomalies in the $d(T)$ curve up to the temperature T_e, which would point to the transition of the system from the micro-homogeneous to microlaminated state, the existence of which has been repeatedly observed near the liquidus in other works (see Section 1.5).

The repeated heating of the melt from temperatures slightly above T_e to the initial temperature (700°C) and the subsequent cooling also yielded the linear temperature dependences of density, which coincide with the curve shown in Fig. 1.22 within the measurement accuracy. Isothermal holding of the specimen at temperatures between T_e and 700°C during 1.5–2 hours revealed no density changes with time, which could also attest to progressive microlamination. Thus, no experimental evidence was obtained in favour of the transition from the homogeneous to microheterogeneous state of the system in cooling the specimens.

The situation changed radically after solidifying and remelting the specimen. In this case anomalies in the temperature dependences of density and the coefficient of thermal expansion of the irradiated zone of the melt were recorded in the course of subsequent heating (Fig. 1.22). These anomalies are similar in appearance to those observed earlier in the temperature dependences of other properties of eutectic melts, which allowed the authors to identify them as signs of the transition of the system from a microlaminated to a homogeneous state. Upon passing through the temperature of anomaly the density values approached those obtained in the course of initial cooling and coincided with them at subsequent thermal cycling of the specimen in the range between 700°C and the eutectic temperature.

Thus, it was found that signs of microlamination of the melt are observed only after melting the eutectic specimen. However, if the melt was brought into the micro-homogeneous state by heating it to a high temperature, the drop in its temperature down to T_e does not result in recovery of the quasieutectic structure.

The retention of the microhomogeneous state on cooling the melt from a high temperature could take place in two cases: if the state of the microhomogeneous solution is thermodynamically stable at all temperatures $T > T_e$ or if this state becomes metastable as the temperature approaches T_e but its lifetime exceeds the duration of experiments. In the latter case the transition of the melt into the microheterogeneous state could be provoked by introducing from outside an inhomogeneity of sufficient dimensions. Special experiments in which weighed portions of lead were introduced at 250°C into the melt preheated to 700°C gave a negative result: no significant effects except for the response of the ordinary system to the addition was noted.

Popel *et al.* [178] concluded from their experiments that the state of a true solution is a thermodynamically stable state of eutectic melts at all temperatures $T > T_e$. The micro-heterogeneity observed after melting the ingot is due to the prolonged retention in the liquid phase of microheterogeneities which are always present in solidified eutectic alloys.

The results of sedimentation experiments in the gravitational field performed by Gavrilin [179] were published simultaneously with the data presented above. The author established that after mixing the components of liquid eutectics (including Sn–Pb) at temperatures of 400–500°C above the eutectic temperature, no indications of sedimentation of

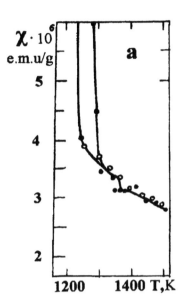

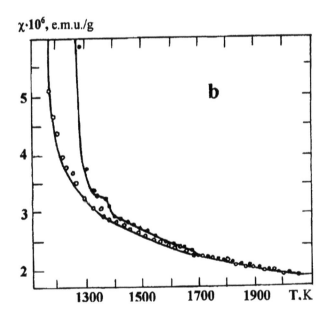

Figure 1.23   Temperature dependences of magnetic susceptibility of Au–27 at. % Co alloy obtained in heating to 1240°C (a) and to 1790°C (b) [180]. The magnetic field strength H = 5.7 · 10^5 A/m.

microsegregations of the heavier component at $T = T_e + 50°C$ were observed. If, however, such a melt is solidified and then remelted, this effect is clearly pronounced in all the systems studied, which supports the hypothesis about the hereditary origin of micro-heterogeneity of liquid eutectics.

Quite indicative results in favour of this thesis were also obtained in [180]. The authors investigated the temperature dependences of the magnetic susceptibility χ of Au–Co melts of eutectic composition. This system is of special interest because the eutectic point lies slightly lower than the Curie temperature T^c of Co-rich solid solutions. Hence, if disperse fragments of the initial eutectic phases are actually retained in the system upon melting, one can expect significant magnetic effects related to the disappearance of ferromagnetism in cobalt-based disperse particles. The results of experiments presented in Fig. 1.23 confirmed this assumption: in the $\chi(T)$ curves obtained on heating the specimen after melting, a characteristic feature is clearly seen near a temperature of 1122°C, coincident with the Curie point of cobalt. Unless the melt was overheated well above T^c, the dependence $\chi(T)$ obtained in its cooling reproduced the heating curve with the indicated anomaly. However, if the hereditary microheterogeneity was destroyed as a result of heating to 1800°C and the system changed over to the state of a true solution, the temperature dependence of susceptibility below 1400°C deviated from the heating curve and had no peculiarities up to the temperature of the onset of solidification of the specimen.

Another important effect, which was confirmed in subsequent studies of other systems, was strongly pronounced in these experiments: after the irreversible transition of the melt into the homogeneous state the specimen solidifies at much deeper supercooling than the non-homogenized specimen. The level of overheating at the crystallization front is known to be the major factor determining the crystal growth and, therefore, the structure and properties of the ingot being formed. As a consequence, the homogenizing overheating of the melt should strongly affect the quality of cast metals.

The most objective argument in favour of the idea of microlamination of eutectic melts formulated in [178] is the results of electron diffraction studies of a Sn–Pb alloy of eutectic composition presented in [181]. After melting the specimen, the maxima of the radial atomic distribution function (RADF) obtained by the Fourier transformation of the structure factor (Fig. 1.24) coincide with the characteristic interatomic spacings in liquid lead and tin. This finding supports the microheterogeneity of the melt. At the temperature 480°C corresponding to the anomaly in the temperature dependence of density presented in Fig. 1.22 the maxima of the structure factor and the RADF are strongly distorted and change their positions, which points to the disappearance of microscopic regions enriched with various components and to the formation of a true solution. As the temperature of the melt drops after its heating above 480°C, the diffraction curves remain virtually unchanged, i.e. the microhomogeneous state is retained.

It is of interest that the specimens heated in the liquid state to temperatures below 430°C solidified on subsequent cooling into a eutectic structure with a distinct triplet of diffraction rings referring to crystalline lead and tin. However, if the melt heated to 480–580°C changed to the state of a true solution, reflections from lead in the diffraction curves of solidified specimens disappeared and the lines of tin were markedly displaced, indicating the formation of an anomalously supersaturated solid solution of lead in tin. This final state was invariably restored after multiple remelting of the resulting solid solution. It was also retained after prolonged holding of the specimen at room temperature. Thus, the first experimental evidence was obtained in this study that the transition of the microheterogeneous melt to the true solution state (this process is called from here on the melt homogenization) is accompanied by dramatic changes in the structure of solidified specimens.

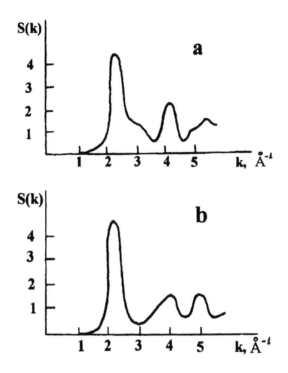

Figure 1.24 Structure factors of Sn–26.1 at. % Pb melt upon melting (T = 190°C, curve (a)) and after heating to 480°C (curve (b)) [181].

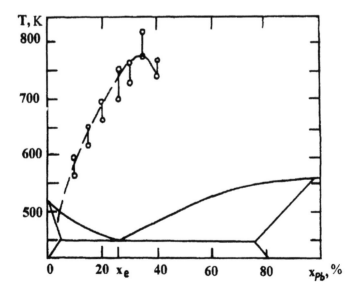

Figure 1.25 Boundary of existence of metastable microheterogeneity on the phase diagram of the Sn–Pb system constructed from the results of measurements of melt density d [182]; vertical bars mark the anomalous portions of $d(T)$ curves.

More recently, Popel *et al.* [178] established that the above-mentioned peculiarities in the behaviour of the temperature dependences of density are also observed in Sn–Pb melts in the case when their composition differs from the eutectic one [182]. Assuming that the anomalous change of the thermal expansion coefficient of these alloys with increasing temperature above the melting point corresponds to the onset of destruction of the micro-laminated state, they constructed the region of stability of such a structure in the region of the phase diagram above the liquidus line (Fig. 1.25). We should emphasize that the boundary of this structure cannot be considered as a line of phase equilibria because it restricts only the zone of stability of microheterogeneity determined by the past history of the melt. The transition across this line after irreversible destruction of the stability zone is not accompanied by a change in the structural state of the system.

Popel *et al.* [182] also revealed the signs of the transition of Sn–Pb melts into the micro-heterogeneous state as the content of the second component increased isothermally to a value corresponding to the line of intersection of the isotherm with the dome-shaped curve plotted in Fig. 1.25. Two series of measurements were carried out. In the first the introduction of successive weighed portions of lead was accomplished at 700°C. The resulting isotherms of density and molar volume are plotted in Fig. 1.26a. Throughout the entire range of concentrations, the $v_\mu(x_{Pb})$ curve shows positive deviations from the additive dependence inherent in regular solutions, and these deviations increased consistently with the growth of x_{Pb}. This curve has therefore no features that could be attributed to the onset of microlamination of the system. The isotherm of molar volume, which was calculated from the concentration dependence of density obtained at 230°C (Fig. 1.26b), shows up in quite a different way. Although the $d(T)$ curve remains monotonic in this case, the isotherm of molar volume exhibits growing positive deviations from the additive straight line only in the initial portion of the curve (approximately up to 10% Pb, i.e. to the point of intersection of the line $T = 230°C$ with the boundary of stability of the microheterogeneous state shown in Fig. 1.25). Then the isotherm of molar volume approaches the additive straight line and, upon crossing it near the eutectic composition, increasingly deviates from it in the opposite direction. Such strange, at first sight, behaviour of the molar volume with a monotonic increase in density was explained by the enhanced microsegregation of the system into Sn- and Pb-rich regions with the growth of x_{Pb} and by sedimentation of these elements in the gravitational field. In this case the molar volume calculated from the mean lead content in the system does not reflect actual conditions in the lower part of the crucible irradiated by gamma-ray photons, because x_{Pb} in it differs markedly from the calculated value.

Long retention in the liquid phase of microheterogeneities inherited from the initial heterogeneous ingot, or introduced in it by isothermal alloying, is possible only with an excess free energy at their boundaries, which is an analog of the ordinary interphase tension between liquids with limited miscibility. This gave the authors of [178, 182] grounds to consider the microheterogeneous eutectic melts as colloidal systems – microemulsions of droplets or microsuspensions of solid particles enriched with one of the components and suspended in a medium of different composition. As opposed to the previously considered colloidal models of liquid eutectics, such a system is assumed to be metastable [178, 182], with a long lifetime, rather than thermodynamically stable. Upon overheating, the melt changes irreversibly to an equilibrium homogeneous state, which is retained at any temperature in the region above the liquidus.

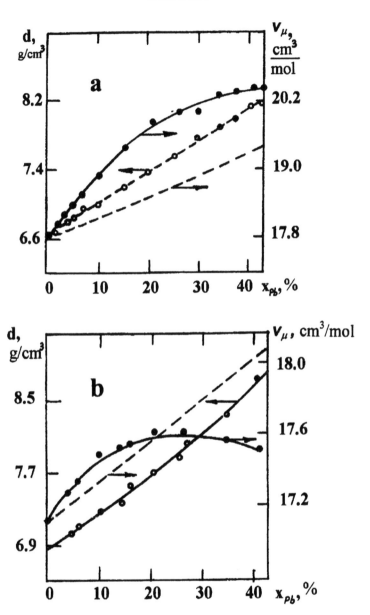

Figure 1.26   Isotherms of density *d* of the irradiated zone of Sn–Pb melt obtained in the course of isothermal alloying of the specimen at 700°C (a) and 230°C (b) [182] and isotherms of molar volume v_μ calculated from these data. Dashed line shows the additive dependence of molar volume on the lead concentration.

From the viewpoint of the colloidal concept, the thermal effects of mixing the components outside the region of microheterogeneity and inside it should be fundamentally different. In the first case the homogeneous solution forms upon introduction of a weighed portion of the second component into the initial melt and the latent heat of dissolving is

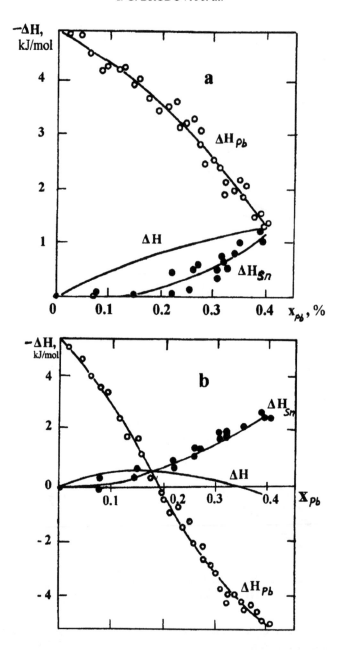

Figure 1.27 Concentration dependences of heats of formation of liquid Sn–Pb alloys at 800°C (a) and 240°C (b).

released or absorbed. In the second case, along with a partial mutual solution of the components, dispergation occurs. The consumption of energy for the formation of interphase boundaries will change the value and, possibly, the sign of the thermal effect of mixing.

The validity of this assumption was supported by the results of measurements of the concentration dependences of the partial (ΔH_i) and integral (ΔH) heats of formation of liquid binary Sn–Pb alloys, in the region of existence of microheterogeneity and outside it. It was found that the total heats of the alloy formation are negative at 800°C (in the region of homogeneous states of the system) and increase in magnitude with the growth of lead concentration in the compositional range studied in the work (up to 40%, see Fig. 1.27a). The sign of ΔH points to the higher energy of interaction of unlike atoms compared to the interaction between like atoms, i.e. gives evidence of the absence of thermodynamical prerequisites for spontaneous formation of microregions enriched with different components. At 340°C the concentration dependence of the total heat of formation is non-monotonic (Fig. 1.27b): upon reaching its peak, ΔH tends to zero and goes over into the region of positive values. The significant deviation of the isotherm from the similar dependence obtained at a high temperature begins at a lead concentration of about 10%. This composition is close to the abscissa of the point of intersection of the straight line $T = 340°C$ with the boundary of the region of metastability of the colloidal state presented in Fig. 1.25, which enables us to relate this effect to the energy expended on the formation of interphase boundaries in the colloid. The increased discrepancy between the high- and low-temperature isotherms of the heat of alloy formation is accounted for by the increased dispersity of colloidal Pb-rich particles as we go deeper into the microlaminated region.

Assuming that the divergence between the temperature dependences of the density in the irradiated zone of Sn–Pb alloys, obtained on heating and subsequent cooling of the specimens (Fig. 1.22), is related to the gravitational sedimentation of disperse Pb particles in the Sn-rich dispersion medium, Popel *et al.* [183] estimated the volume fraction Φ and the mean radius R of such particles. The results of these calculations are presented in Table 1.3.

Independent estimates of the size of disperse particles in Sn–Pb melts were made using the results of the study of the heat of mixing. By relating the positive deviations of ΔH in

Table 1.3 Radius R, nm (numerator) and volume fraction Φ of fine particles (denominator) at various temperatures and compositions.

T, °C		R (nm) and Φ at alloy composition x, at. %					
	10	15	20	26.1	30	35	40
200	11.3 / 0.08	10.6 / 0.16	9.5 / 0.23	9.0 / 0.31	9.5 / 0.37	8.3 / 0.44	10.0 / 0.51
250	13.2 / 0.06	11.4 / 0.14	11.0 / 0.22	9.5 / 0.32	10.4 / 0.38	8.9 / 0.46	10.4 / 0.54
300	14.4 / 0.04	12.4 / 0.13	11.6 / 0.22	10.2 / 0.33	11.0 / 0.40	10.2 / 0.49	10.7 / 0.58
350		13.2 / 0.05	11.4 / 0.17	11.2 / 0.32	12.7 / 0.41	12.9 / 0.53	12.4 / 0.65
400			14.1 / 0.21		14.9 / 0.35	14.9 / 0.52	14.9 / 0.69
450					17.4 / 0.29	17.4 / 0.64	

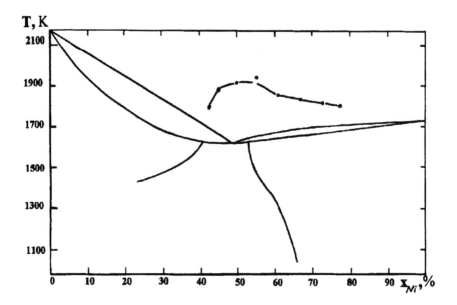

Figure 1.28   Region of existence of metastable microheterogeneity on the phase diagram of the Ni–Cr system constructed from the results of densitometric study [184].

the microlamination region from the values characteristic of homogeneous solutions to the energy expended on the formation of interphase boundaries in the colloid, and estimating the interphase tension from Antonov's rule (see Section 1.5), Popel *et al.* [183] obtained R values ranging from 110 to 3–4 nm, which decrease regularly as we go isothermally deeper into the microheterogeneity region with increasing lead concentration. These results, as well as those presented in Table 1.3, are consistent with the estimates derived from sedimentation experiments [153–156, 179].

After the idea of metastable microheterogeneity of melts had been verified experimentally, the authors turned to a study of specific manifestations of this effect in melts with different kinds or even different types of phase diagrams.

The temperature dependences of density of melts of the Ni–Cr and Fe–Cr systems were examined by Makeev and Popel [184]. The first system has a simple eutectic phase diagram with a very extensive region of solid solutions at high temperatures. (Near the eutectic temperature the length of the concentration interval, where the heterogeneous eutectic structure exists, does not exceed 12–13%.) The second system falls in the class of systems with a continuous series of solid solutions.

In experiments with Ni–Cr melts, as well as in the above-mentioned studies, the branching of $d(T)$ curves obtained on heating the specimen from the room temperature and subsequently cooling was revealed. The authors related it to the irreversible destruction of the hereditary microheterogeneity. The boundary of the region of existence of the colloidal state, which was constructed with the use of branching points (Fig. 1.28), shows that in this case the microheterogeneity persists upon melting only in a comparatively narrow temperature–concentration range. In [184] evidence was also obtained showing that the microheterogeneous state is really metastable, and is not a non-equilibrium state with prolonged relaxation time. The authors carried out a series of experiments in which they measured the relaxation times of the system after mixing its components at various temperatures. In cases where the temperature of mixing was 40–50°C lower than the temperature of transition of the melt to the homogeneous state, they observed a rapid (within a few minutes) change in density and its approach to the values obtained on cooling from the homogeneous state after the initial relaxation of density of the system had been attained and stable density values had been recorded over the 40–60 min period. This threshold relaxation regime is characteristic of the destruction of metastable states. If the microheterogeneous state had been non-equilibrium, an exponential density change with time, typical of relaxation, would have been observed.

In some experiments, Makeev and Popel [184] after solidifying the specimen homogenized in the liquid state lowered the temperature only to 1000°C and then remelted the specimen and recorded the $d(T)$ dependences on heating and cooling. No branching of these curves was observed in this case, which led the authors to a conclusion about the radical change of the way in which solidification of homogenized specimens occurs as compared to microheterogeneous samples. It is evident that in systems with an extensive region of solid solutions the driving force of the eutectic transformation is small. For this reason, in the absence in the melt of regions enriched with various components, a strong supercooling of the system is necessary to separate it into two phases. The homogenized melt solidifies initially in the form of an anomalously oversaturated solid solution, which then undergoes a transformation of a eutectoid type at a lower temperature. If before the onset of this transformation the specimen is repeatedly heated and melted, the initial homogeneity of the melt is retained.

The study of the temperature dependences of density of Fe–Cr melts showed that no specific features were observed in these curves after the system's relaxation related to the mixing of its components had been completed. Solidification and remelting of the relaxed melt also do not lead to the appearance in the $d(T)$ curves of anomalies that might be attributed to the destruction of microheterogeneity. This finding led the authors of [184] to the conclusion that in melts of systems with a continuous series of solid solutions the micro-heterogeneous states arising during the mixing of components are non-equilibrium, and on completion of the relaxation process the melt reaches the state of a true solution at any temperature above the liquidus. Since the solidification of such systems is not accompanied by their separation into two phases, the homogeneity of the specimen is retained both in the solid state and after remelting. Consequently, there is no need for significant overheating of the specimens above the liquidus to reach the homogeneous state of melts of this type: it suffices to hold the melt at the usual temperature of melting until the equilibrium is attained, bearing in mind that near the liquidus the relaxation process usually lasts a few hours.

The data presented here, in combination with the results of several subsequent studies, indicate that the tendency of systems toward the formation of a metastable colloid correlates with the mutual solubility of the components in the solid state. Even though the microheterogeneous state arises in melts of the systems with a continuous series of solid solutions, it proves to be non-equilibrium and relaxation ends in the equilibrium homo-geneous state. In eutectic systems with significant mutual solubility of the components in the solid state, metastable microheterogeneity exists in comparatively narrow limits of temperatures and compositions. And finally, in liquid eutectics with low mutual solubility

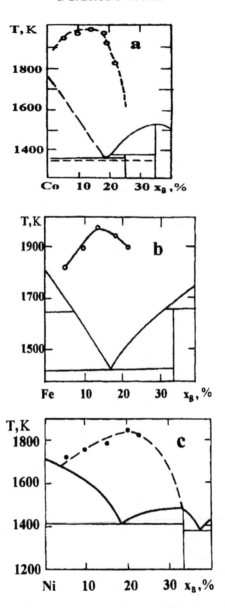

Figure 1.29   Boundaries of the regions of metastable microheterogeneity of melts on the phase diagrams of the Co–B, Fe–B, and Ni–B systems, according to the data of [186–188].

in the solid state the region of existence of the colloid covers an extensive region of the phase diagram.

The specific features of realization of the microheterogeneous state in eutectic systems with compounds have been studied in a series of works. The authors investigated the temperature dependences of density [185] and surface tension [186] of Ni–B melts, density and resistivity of Fe–B [187] and Co–B melts [188], as well as the density of Fe–Sc, Fe–Ge, and some other objects. In the systems with compounds, indications of microheterogeneity were also revealed upon melting the specimens or mixing the components at insignificant overheating above the liquidus line. This microheterogeneity is due to the prolonged retention of inhomogeneities inherited from the initial crystalline specimens. Heating of the melts to definite temperatures is accompanied by the irreversible destruction of this state, which is evidenced by the branching of the temperature dependences of their properties. The branching points were used to construct the regions of existence of metastable heterogeneity on the phase diagrams of the Fe–B, Ni–B, and Co–B systems (Fig. 1.29).

Assuming that the abscissas of the points of intersection of the dome-shaped boundaries of these regions with the corresponding isotherms determine the compositions of contacting phases, the authors of [187, 188] calculated the interphase tension σ at the surface of disperse particles in Fe–B and Co–B melts. The resulting σ values are sufficiently low (lie in the range between 5 and 21 mJ/m^2) to cause at temperatures above the liquidus a spontaneous dispergation of boride phase fragments inherited from the initial crystalline specimens. For this reason, the authors estimated the dispersity of colloidal particles in these melts using Rebinder's criterion which relates the minimum size R of particles in such a dispergation to the tension σ at the interface between the phases:

$$\sigma \approx \gamma k T / R^2. \qquad (1.21)$$

Here k is the Boltzmann constant and $\gamma \approx 20$–30. The resulting R values fall in the range 6–13 nm, which agrees with the estimates of radii of colloidal particles in simple eutectics made above.

New features in the temperature dependences of properties, which have not been revealed in experiments with simple eutectics, were observed in the study of compound-containing systems. In particular, anomalies near the temperatures of melting of intermetallic compounds were found in the $d(T)$ curves of Fe–Sc and Fe–Ge specimens on heating. If the cooling of the specimen began before attaining the homogeneous state, these anomalies were also reproduced in the cooling curves. However, in the cases where the melt was transformed to the true solution state, smooth curves with no anomalies were obtained on cooling. This led the researchers to a radical hypothesis on the possibility of retaining the crystalline structure of disperse particles in the region above the liquidus and on the melting of these particles near the corresponding temperatures. If these ideas are correct, polymorphic transformations in these particles can also be expected at lower temperatures, including transformations that are absent on the equilibrium phase diagrams of compounds. The reasons for such transformations may be either high capillary pressure inside particles, which is caused by the interphase tension ($p = 2\sigma/R$), or the size factor (large ratio of the surface energy to the volume energy).

Later, the temperature dependences of density of high-purity Fe–B melts were examined in [59] In this case, in contrast to the results obtained in [187], indications of the transition of the system to the homogeneous state were detected at higher temperatures. On the one hand, this gives an additional verification of the colloidal nature of microheterogeneity of melts, because impurities present in the previously investigated specimens could only

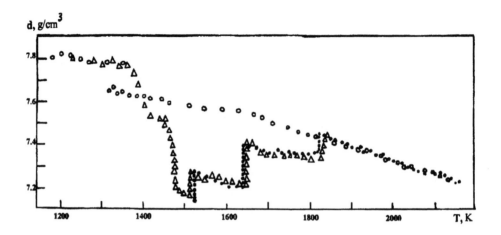

Figure 1.30   Temperature dependences of the density of Ni–22.5 at. % B alloy [185] (• first heating, ○ cooling, ▵ repeated heating after solidification).

reduce the interphase tension at the boundaries of disperse particles, which determines their thermal stability. On the other hand, the preservation of the colloidal state at higher temperatures in the experiments described in [59] made it possible to reveal new features in the $d(T)$ curves at 1650–1750°C, i.e. well above the melting point of the nearest inter-metallic compound Fe_2B. Since this anomaly was observed most distinctly in the specimen of stoichiometric composition, the authors related it to a rearrangement of the structure of short-range order in disperse particles of the molten compound (effect discussed in Section 1.5).

The temperature dependences of density of Ni–B melts had an even more unusual appearance [185]. Stable values of density in the irradiated zone of the melt, decreasing regularly with an increase in temperature, were initially detected on heating the melt after melting the specimen and attaining the equilibrium state (Fig. 1.30). Then, the density instability was observed at several (2 or 3) "critical" temperatures $T_{(c)}$ and over 5–10 hours the density varied with time. After a new "equilibrium" density had been established, a stable linear dependence $d(T)$ was obtained again for a certain temperature range, and upon attaining the next "critical" temperature an isothermal change in density was repeatedly observed with emergence onto a new "equilibrium" level. On completion of the relaxation process at the last "critical" temperature $T_{(c)}$, the heating curve became and remained linear in subsequent thermal cycling of the melt. If the specimen was solidified and remelted, the same features as in the first heating were revealed.

The results obtained show that the transition of a microheterogeneous melt to the state of a true solution is achieved in some cases after a few successive changes in the colloidal structure, which cause the density of the irradiated zone to "drift" upon attaining "critical" temperatures. These changes may be related to the dispergation of colloidal particles to a size allowing the retention of their metastable state up to the next critical temperature T_{cr}. The possibility of such transitions between two two-phase states was verified theoretically by Pushl and Abauer [189]. Obviously, in each intermediate transition the composition of the dispersive medium and disperse particles changes and a more uniform distribution of particles over the specimen height is established.

The Fe–B, Ni–B, and Co–B systems constitute the basis of amorphous alloys which are usually obtained by the method of quenching from the liquid state and possess a unique set of service characteristics. A large body of data on microlamination of amorphous ribbons accumulated over the time of their study may be explained by the existence of metastable microheterogeneity of the melts of these alloys. The first indirect data in this context were obtained in studies of magnetic susceptibility [190] and Hall effect [191] of Pd–Si based specimens, and saturation magnetization of Co–B amorphous alloys [192]. The use of the methods of electron and field ion microscopy, small-angle scattering of X-rays and neutrons made it possible to estimate the sizes of precipitates of the disperse phase. A brief summary of the results is given in Table 1.4. The experimentally determined dispersity scale of the structure of metallic glasses agrees reasonably well with the estimates of the size of colloidal particles in Fe–B and Co–B melts presented above. This fact suggests, on the one hand, that the microheterogeneity of initial melts is inherited by amorphous ribbons; on the other hand, it gives grounds to use the results of investigation of the disperse phase of metallic glasses for gaining additional information on the colloidal states of liquid

Table 1.4   Sizes of precipitates of disperse phase in metallic glasses according to the data of various researchers.

System	Method of investigation	Size of precipitates, nm	Reference
Fe–B	SANS	4.3–4.6	[193]
	SANS	~1	[194]
Fe–B–Si	SANS	4.0	[193]
Fe–Al	SANS	6–15	[195]
Fe–Zr	SAXS and EXAFS	0.6–1.2	[196]
Fe–P–C	SAXS	1.8–2.4	[197]
	SAD	~10^3	[198]
Fe–Ni–B	SANS	~1	[194]
Fe–Ni–P–C	SANS	160	[199]
	HREM	25	[200]
Fe–Ni–Mo–B	SAD	~10^3	[198]
Fe–Ni–Cr–Mo–B	SAD	~10^3	[198]
Fe–Cr–P–C	SAD	~10^3	[198]
Ni–Nb	FIM	3–5 (580°C)	[201]
		10–15 (650°C)	
Ni–B	SAXS and SANS	47	[202]
Tb–Cu	SANS	~100	[203]
Pd–B	SAXS	5 and 15	[204]
Pr–Ga–Fe	SAD and EPMA	40	[205]

Note: SANS – small-angle neutron scattering; SAXS – small-angle X-ray scattering; EXAFS – extended X-ray fine structure spectroscopy; SAD – selected area diffraction; HREM – high-resolution electron microscopy; FIM – field ion microscopy; EPMA – electron probe microanalysis.

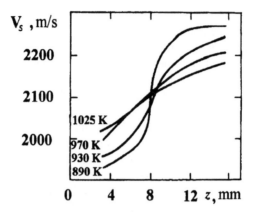

Figure 1.31   Ultrasound speed V_s as a function of the height level z in the Ga–30 at. % Pb melt at different temperatures [186].

alloys. According to the data obtained from the study of X-ray and neutron selected area diffraction patterns (SADPs) [198], Hall effect [206], saturation magnetization [207], as well as data from field ion microscopy [208] and EXAFS spectroscopy [209], one of the phases of the microheterogeneous amorphous structure originates from the α solid solution, whereas the second comes from the compound nearest to this solution. The sizes of disperse inclusions increase with temperature. Boucher *et al.* [203] observed adsorption of impurities on the internal interphase boundaries of an amorphous $TbCu_{3.54}$ alloy. Nowadays microsegregation in metallic glasses is considered to be a well-established fact and attempts are being undertaken to construct model theories of the amorphous state based on the concept of their two-phase structure [210].

The presence of disperse inclusions and the change in their size and composition with annealing have a significant effect on the properties of metallic glasses. For example, Osamura *et al.* [211] detected an increase in the ultimate strength and fracture toughness, while Haasen [212] observed embrittlement of amorphous alloys caused by the increase in the microheterogeneity scale on heating. A decrease in the thermal conductivity caused by the phonon scattering from disperse Fe_3P inclusions was revealed by Pompe *et al.* [213]. According to [191], microsegregation plays a key role in forming magnetic properties of metallic glasses. It was also shown in [214] that particles of the disperse phase serve as crystallization centres and that the thermal stability of amorphous alloys depends on the composition and dispersity of these particles. Clearly, changing the particle size in the colloidal state of the melt by varying its temperature or by means of other external actions may have a significant influence on the structure and properties of metallic glasses.

The concept of metastable colloidal states can also be extended to the melts with a miscibility gap. As the temperature of such a system rises, a macroscopically inhomogeneous melt forms initially with a distinct interface between immiscible liquids. With further heating and passing through the lamination dome, the system should change to a true solution state. However, the initial heterogeneity gives rise to the dispergation of one phase within the other to the nanometer sizes; the resulting droplets of the disperse phase exist for a long time in the melt forming a metastable emulsion. This is evidenced by the results of the ultrasonic study of Ga–Pb melts performed by Philippov and Popel [186]. They measured the speed of ultrasound V_s in this system using the pulse-phase method with variable acoustic base. This method allows the determination of V_s on various height levels z of the melt. The authors observed a significant dependence of the speed of ultrasound on the vertical coordinate (Fig. 1.31), which was attributed to the compositional inhomogeneity over the specimen height, i.e. to the sedimentation of heavy disperse particles enriched with Pb in the field of natural gravity. This effect appears after mixing the components at temperatures 20–200°C above the dome of macroscopic lamination or on heating the specimen through and beyond the immiscibility region. The effect persisted throughout the experiment (during ~10 hours). However, if the melt was overheated above the dome by 300–400°C, no signs of its inhomogeneity with height were found in subsequent cooling, which points to the metastability of the microheterogeneous state. It is of interest that the subsequent process of macroscopic decomposition of the system into two liquids in the immiscibility region, i.e. its transition to the true solution state, also took a much longer time, thereby affecting markedly (as in eutectic systems) the course of phase transformations in cooling. Using the results of measurements of the speed of ultrasound at various distances from the crucible bottom, Popel and Sidorov [186] determined the dependence of the melt composition on this distance and, assuming Boltzmann distribution of Pb-rich particles in the gravitational field, estimated the radius R of these particles. The resulting R values fall in the range 17–23 nm, i.e. disperse particles in liquid Ga–Pb alloys have approximately the same dimensions as in melts of the eutectic type.

Thus, the considerable experimental evidence accumulated to date shows that liquid metallic solutions may exist for a long time in a state of metastable heterogeneity inherited from the initial crystalline materials or originated in mixing the components at low overheating above the liquidus.

From this point of view, three, at first glance dissimilar, phenomena mentioned at the beginning of this section can be interpreted. In particular, the structural metallurgical heredity is related to the dependence of the size and composition of disperse particles in the melt on the scale of heterogeneity of initial materials and to the influence of these factors on the solidification process. The predisposition of eutectic melts to microsegregation, with the formation of regions enriched with various components, is accounted for by the heterogeneity of the crystalline structure of eutectics. Finally, the hysteresis of temperature dependences of properties of the melt is associated with irreversible destruction of the colloidal structure and with transition of the system to a true solution state. An understanding of the nature of these phenomena provides a reliable basis for searching for different ways of action on the structure of microheterogeneous melts with the aim of improving the structure and properties of ingots formed by solidification of melts.

1.8   Thermodynamical Basis for the Concept of Metastable Microheterogeneity of Melts

The ideas about the possibility of anomalously slow solution of fine particles inherited from initial crystalline materials, or their state of metastable equilibrium within the surrounding melt in systems with unlimited miscibility of components in the liquid state, called for theoretical substantiation. The first attempt to explore this problem by thermodynamical

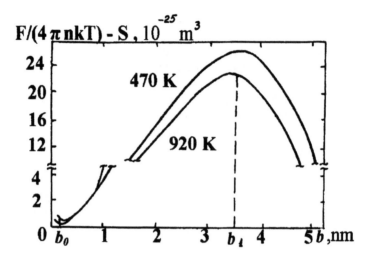

Figure 1.32 $F/(4\pi nkT) - S$ versus the thickness b of the transition layer for the particle of radius $R = 10$ nm in the Sn–Pb melt at different temperatures ($x_1 = 0.99999$, $x_2 = 0.00001$).

methods was undertaken by Popel *et al.* [169]. The authors analyzed a model system – a spherical particle of radius R and composition x_1 in a dispersive medium with characteristic size L and composition x_2. To the first approximation, the concentration profile in the transition layer of thickness $2b$ was assumed to be linear. The free energy F of such a system was calculated using the gradient approximation of thermodynamics of inhomogeneous systems (the Cahn–Hilliard approximation [215]) using the model of regular solutions. As a result, the following expression was obtained:

$$F = 4\pi nkT(KR^3 + MR^2b + NRb^2 + Pb^3 + Q(R^2/b + b/3) + S), \qquad (1.22)$$

Here n is the atomic number density and T is the temperature. The coefficients K, M, N, and P depend on x_1, x_2, and on the ratio of the interchange energy ω (which specifies in the regular solution model the energetic non-equivalency of interactions of like and unlike atoms) to the mean energy of their thermal motion kT. The Q factor depends, in addition, on the mean radius of action of the interatomic potential, while S is the function of the system size L. Analytic expressions for these coefficients are given in [216]. With the fixed values of the above parameters, the free energy is determined only by the radius of the particle R and the thickness of the transition layer b.

More recently calculations of the free energy of the given model system were also carried out using a more realistic, exponential concentration profile in the transition layer. In this case the integration of the expression for free energy can be performed by numerical methods. The results of these calculations agree well with (1.22). For this reason, in the following discussion we can restrict ourselves to analysis of equation (1.22). To be specific, all the calculations in [169] were carried out for the Sn–Pb system. The temperature and concentration boundaries of existence of microheterogeneity in this system have been determined in the densitometric experiments [182]. The energy ω was set to be

$8.31 \cdot 10^{-19}$ J/atom. This value was calculated from the isotherm of the heat of formation of Sn–Pb alloys at 800°C. The characteristic radius R_φ of action of the interatomic potential was assumed to be 0.5 nm. This estimate is rather arbitrary, but the results of the calculation of the free energy in the Cahn–Hilliard approximation show a weak sensitivity to the accuracy of determination of the gradient term in the expression for free energy density and to the R_φ value appearing in this term.

The typical dependence of the quantity $F/(4\pi nkT) - S$ on the thickness of transition layer b for the particle with radius 10 nm is presented in Fig. 1.32. When $b \to \infty$, the free energy of the system tends to an absolute minimum. This implies that the true solution state is a thermodynamically stable state for liquids with unlimited mutual solubility of the components. However, an additional minimum of F exists at the thickness b_0 of the transition layer, and a maximum exists at b_1. Such a non-monotonic dependence $F(b)$ was accounted for in [169] by the competitive influence of two processes: the decrease in the free energy with increasing the thickness of the transition layer due to the decrease in the concentration gradient and, second, the growth of F caused by the involvement of new particles in this layer.

If a particle of radius R with the thickness of the transition layer $b < b_1$ is embedded in the melt, it will not be dissolved in it in a diffusive way with increasing δ, but will change to a state with the "equilibrium" value $b = b_0$. Then, in general, it continues to increase or, conversely, to diminish in size. However, the rate of this process is no longer limited by diffusion in the volumes of contacting phases but rather by the transfer of particles across the energy barrier at the boundary created by the gradient term in the expression for the local density of free energy. We call this solution regime the process of kinetic dissolving.

Since the transition layer of thickness b_0 forms in a time shorter than the time of a significant change in the particle radius R, the process of kinetic solution of a droplet can be considered as quasi-equilibrium. This justifies the use of methods of equilibrium thermodynamics for determining the b_0 values at various concentrations x_1, x_2, and radii R. The calculations show that if we take the values that were determined in [182] from the positions of the boundaries of the region of metastable microheterogeneity in Sn–Pb melts as the compositions x_1 and x_2 inside and outside the particle then the temperature rise at a fixed R will be accompanied by a weak change in the "equilibrium" thickness b_0 of the transition layer. However, the local minimum between 400 and 450°C in the $F(b)$ curve disappears and the transition of dissolving into the kinetic regime becomes impossible. According to [182], indications of the microheterogeneity of Sn–Pb melts are observed only at $T < (450–470)$°C. Therefore, the prolonged retention of disperse particles of colloidal size is due to the implementation of the kinetic regime of their solution.

At fixed compositions of the contacting phases, the "equilibrium" thickness b_0 of the transition layer grows approximately linearly with particle radius and depends weakly on temperature (Fig. 1.33). Significant deviations from the linear dependence are observed only for $b_0 < 0.3$ nm, i.e. in cases when the characteristic scale of the compositional inhomogeneity is smaller than the effective radius R_φ of action of the interatomic potential and the gradient approximation of the free energy is inapplicable. A decrease in the thickness of the transition layer with reduction in the particle radius leads to growth of the concentration gradient, which determines the height of the energy barrier at the boundary and the rate of kinetic mass transfer. For particles with radius less than 10 nm, the b_0 values differ only slightly from the interatomic distance. Therefore, they have a monoatomic

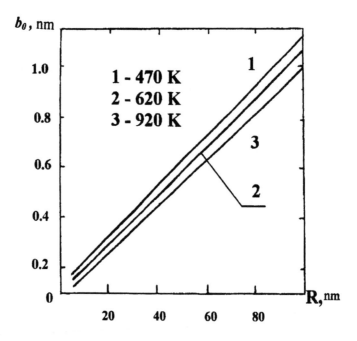

Figure 1.33 "Equilibrium" thickness b_0 of the transition layer versus the particle radius at different temperatures ($x_1 = 0.99999$, $x_2 = 0.00001$).

transition layer, similar to that formed between immiscible liquids away from the critical point of lamination, and possess excess free energy of the same order of magnitude.

The specific excess free energy σ of the transition layer was determined as the difference between the free energies F per unit area of the interphase surface for the system with the model concentration profile and for that consisting of homogeneous regions with the compositions x_1 at $r < R$ and x_2 at $r > R$ (in the latter case the boundary does not contribute to F). A sphere of radius R corresponding to the local concentration $x = (x_1 + x_2)/2$ plays the role of a separating surface. The results of these calculations are presented in Fig. 1.34. For particles with radius more than 10–30 nm the σ value is negative, but it grows and becomes positive as R decreases. If we identify the specific excess free energy of the transition layer with the interphase tension at the boundary, the results of the calculations will attest to the thermodynamical advantage of spontaneous dispergation of large particles enriched with one of the components to a size of ~10 nm. The conditions required for metastable equilibrium of such particles with the dispersion medium were examined by Barboi *et al.* [217]. The authors showed that equilibrium is possible if the interphase tension grows more rapidly than by the $1/R^2$ law. In the case considered here, this condition is satisfied at $R >$ 10 nm (Fig. 1.34). This means that thermodynamical prerequisites exist for realization of metastable colloidal states in melts with unlimited mutual solubility of the components.

The indicated character of variation in the interphase tension with decreasing particle radius takes place at any composition of contacting phases. However, experiments show that the metastable microheterogeneity of melts is retained only in the temperature–concentration region of the phase diagram bounded by the dome-shaped

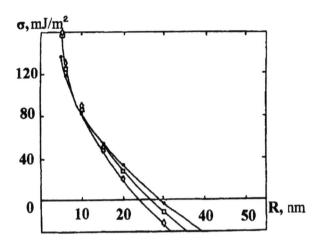

Figure 1.34 Specific free energy σ of the transition layer with "equilibrium" thickness b_0 as a function of the particle radius at different temperatures ($\bullet$ 200°C, $\square$ 450°C, $\diamond$ 650°C).

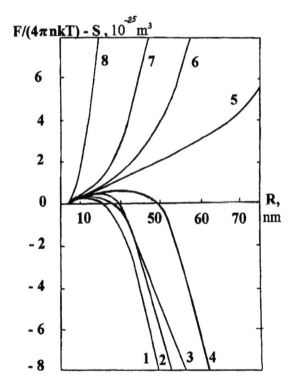

Figure 1.35 Free energy of the system versus the radius of the particle at the boundary of which the "equilibrium" thickness b_0 of the transition layer is kept constant (the concentration of the second component x_2 is 0.00001 (curve 1); 0.001 (curve 2); 0.003 (3); 0.005 (4); 0.006 (5); 0.007 (6); 0.01 (7); 0.1 (8)).

curve shown in Fig. 1.25. According to the results of densitometric and calorimetric experiments, if the weighed portions of the second component are introduced isothermally into the first, molten component, their initial rapid solution is followed by dispergation of successive portions of the second component with the formation of a metastable colloid, which occurs, however, only after a certain threshold concentration has been attained. The reasons for this will become clear if we analyze the relation between the free energy of the system and the radius of the particle, at the boundary of which the "equilibrium" thickness of the transition layer corresponding to the current R value is kept constant. The results of the calculation for the $F(R)_{b_0}$ function are presented in Fig. 1.35. Analysis of these results shows that at low contents of the second component in the dispersive medium a maximum is present in the $F(R)_{b_0}$ curve in the range of particle radii ~10 nm. For larger particles the growth is more advantageous from the thermodynamical point of view than their solution. According to [218], in this case the particle growth rate dR/dt increases with radius and its equilibrium in the dispersive medium is impossible. As the content of the second component in the melt surrounding the particle increases, the maximum of $F(R)_{b_0}$ moves initially toward larger R and then, at a certain threshold concentration x_2^0, disappears completely. Starting from this composition, the process of size reduction becomes thermodynamically advantageous for particles of any radius; it is accompanied by the growth of interphase tension and ends in the state of metastable equilibrium with the dispersive medium. Thus, the abscissa of the point of intersection of the corresponding isotherm with the dome-shaped boundary of the microlamination region on the phase diagram corresponds to the concentration x_2^0.

More recently similar calculations were carried out for Ni–B and Fe–B melts. In these systems the metastable equilibrium of disperse particles in the dispersive medium enriched with the main component is also possible; the corresponding compositions of these particles are close to Ni_3B and Fe_2B. The most likely dimensions of equilibrium particles lie in the ranges 120–200 nm and 10–20 nm, respectively, which agrees reasonably well with the characteristic scales of microheterogeneity of metallic glasses obtained by rapid quenching of these melts (Table 1.4).

The next step toward the theoretical justification of the possibility of metastable equilibrium of disperse particles in melts was the thermodynamical analysis of free energy of the bounded system involving the disperse particle surrounded by a small volume of the dispersive medium. Such a system represents a model of the cell of the colloidal melt per particle. For monodisperse colloid, the sizes of all the cells are on average identical and do not vary with time. This means that no mass transfer occurs at the cell boundary. Therefore, the number of atoms of each component in the system remains constant and any stage of the particle's evolution is accompanied by a change in the free energy due to variations in the volume and composition of the surrounding dispersive medium. From the mathematical viewpoint, the problem of evolution of the disperse particle in such a cell is equivalent to the stability analysis of the nucleus of the new phase in the bounded system, which was performed by Rusanov [219].

Assume that at any stage of the system's evolution the conditions for thermodynamical equilibrium are satisfied in the interior of the disperse particle, i.e. the Gibbs–Duhem equation holds:

$$A\,d\sigma - v^{(1)}dp^{(1)} + S^{(1)}dT^{(1)} + \Sigma N_i^{(1)}d\mu_i^{(1)} = 0. \qquad (1.23)$$

Here, A is the particle's surface area; σ is the interphase tension at the particle's boundary; N_i is the number of atoms of kind i; μ_i is their chemical potential; p, v, and T are the thermodynamical parameters; and S is the entropy; superscript (1) relates to the particle. At constant p and T, the equilibrium corresponds to the minimum of Gibbs thermodynamical potential

$$G = U - TS + pv, \tag{1.24}$$

where U is the internal energy. For a homogeneous system,

$$U_{\text{hom}} = TS - pv + \Sigma \mu_i N_i, \tag{1.25}$$

$$G_{\text{hom}} = U_{\text{hom}} - TS + pv = \Sigma \mu_i N_i. \tag{1.26}$$

For a two-phase system containing a disperse particle:

$$U_{\text{het}} = TS - p^{(1)} v^{(1)} - p^{(2)} v^{(2)} + \sigma A + \Sigma \mu_i^{(1)} N_i^{(1)} + \Sigma \mu_i^{(2)} N_i^{(2)}, \tag{1.27}$$

$$G_{\text{het}} = U_{\text{het}} - TS + p^{(2)} v = (p^{(2)} - p^{(1)}) v^{(1)} + \sigma A + \Sigma \mu_i^{(1)} N_i^{(1)} + \Sigma \mu_i^{(2)} N_i^{(2)}, \tag{1.28}$$

where superscript (2) refers to the dispersive medium.
Following [38], we assume that

$$\mu_i^{(1)}(p^{(1)}) \approx \mu_i^{(1)}(p^{(2)}) + v_i^{(1)}(p^{(1)} - p^{(2)}), \tag{1.29}$$

where $v_i^{(1)}$ is the partial molar volume of the ith component in the particle. Then,

$$G_{\text{het}} \approx \sigma A + \Sigma N_i^{(1)} \mu_i^{(1)}(p^{(2)}) + \Sigma N_i^{(2)} \mu_i^{(2)}. \tag{1.30}$$

At $N_i = N_i^{(1)} + N_i^{(2)} = \text{const}$, the difference between Gibbs potentials of the heterogeneous and homogeneous systems (G_{hom} corresponds to the global minimum of G) is determined by the expression:

$$\Delta G = (p^{(2)} - p^{(1)}) v^{(1)} + \sigma A + \Sigma (\mu_i^{(1)} - \mu_i^{(2)}) N_i^{(1)} + \Sigma (\mu_i^{(2)} - \mu_i) N_i. \tag{1.31}$$

We assume that at any stage of the particle's evolution the condition for mechanical equilibrium of the particle with the surrounding melt is satisfied:

$$p^{(1)} - p^{(2)} = 2\sigma/R \tag{1.32}$$

(R is the particle radius). In the approximation of regular solutions we have

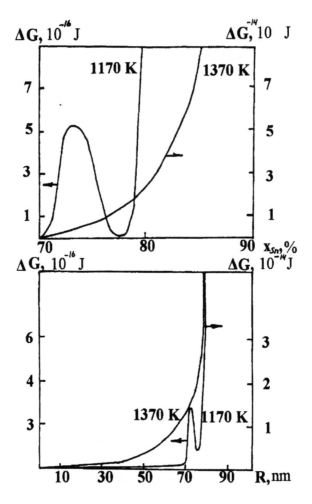

Figure 1.36   The change of the Gibbs thermodynamical potential ΔG of the Al–70 at. % Sn melt caused by the appearance of a disperse particle as a function of composition x_1 and radius r of the particle at different temperatures.

$$\mu_i = \mu_{i_0} + R_g T \ln x_i + \omega_\mu (1 - x_i)^2. \tag{1.33}$$

Here, R_g is the gas constant; x_i is the concentration of the ith component expressed in mole fractions; $\omega_\mu = N_A z \omega$ is the molar interchange energy; z is the coordination number; N_A is Avogadro number; and ω is the interchange energy per one atom.

Then we obtain for a binary system the following expression:

$$\Delta G = (4/3)\pi R^2 \sigma + (R_g T \ln((1 - x^{(1)})/(1 - x^{(2)})) + \omega_\mu((x^{(1)})^2 - (x^{(2)})^2)v^{(1)}(1 - x^{(1)}) +$$

$$(R_g T \ln(x^{(1)}/x^{(2)}) + \omega_\mu(x^{(1)} - x^{(2)})(x^{(1)} + x^{(2)} - 2))v^{(1)}x^{(1)} + \tag{1.34}$$

$$(R_g T \ln (x^{(2)}/x + \omega_\mu (x^{(2)} - x)(x^{(2)} + x - 2)) v x +$$

$$(R_g T \ln ((1 - x^{(2)})/1 - x) + \omega_\mu ((x^{(2)})^2 - x^2)) v(1 - x),$$

where x and v are the concentrations of the second component in mole fractions and the numbers of moles in the phases, respectively.

The local minimum of this expression with respect to two independent variables x_1 and R corresponds to the metastable equilibrium of the particle with the dispersive medium.

Popel *et al.* [169] carried out the numerical minimization of expression (1.34). The interphase tension at each minimization step was calculated by the method described above. The molar energy of mutual exchange corresponded to Al−Sn melts (21 400 J/mole) and the cell size was of the order of 100 nm.

As a result of these calculations, the local minimum with a depth of $(1-5) \cdot 10^{-16}$ J was discovered at the surface $\Delta G(R, x^{(1)})$ at 900°C (Fig. 1.36). It corresponds to the metastable equilibrium of the colloidal particle with the dispersion medium. On heating to 1100°C, this minimum disappears and metastable equilibrium becomes impossible. The indicated temperature is close to the branching points in the temperature dependences of viscosity of Al−Sn melts presented in Section 2.1.

Thus, the concept of metastable microheterogeneity of liquid metallic solutions is not only verified to date by numerous experiments, but is also substantiated from the thermo-dynamic point of view. The results of thermodynamical calculations are in satisfactory agreement with the estimates based on experimental data. This allows us to consider the above concept as a sufficiently rigorous theory of the melt structure, rather than as a more or less plausible hypothesis, and to use the results of this theory for analysis of the efficiency of various methods of external action on liquid metal with the aim of improving the quality of ingots, castings, and amorphous ribbons.

1.9   Perspective Methods of Liquid Aluminium Alloys Processing

The data on structural rearrangements in metallic melts presented in the previous sections made it possible to single out the most promising ways of external action on liquid aluminium alloys with the aim of improving the quality of cast and deformed metals.

The lack of unambiguous evidence for structural transformations in pure liquid aluminium (see Section 1.2) does not permit one to recommend heat−time treatment for this purpose. It is evident that success can be expected only if specific additives are introduced in the molten aluminium to stimulate the formation of impurity clusters influencing significantly the solidification process (this effect was discussed in Section 1.4).

The most challenging way is to use various external energy actions upon liquid aluminium alloys retaining the metastable microheterogeneous state, inherited from initial heterogeneous materials in the melting process in most industrial smelters. Overheating above the liquidus to a temperature higher than the temperature of irreversible transition of the system to the true solution state enables one to influence significantly the conditions of formation of phases in the solidification process and, further, the structure and properties of ingots and castings. Increasing the melting temperature without going beyond the region of metastable microheterogeneity can lead to less significant but equally useful effects. The changes in dispersity and composition of phases of the colloidal melt achieved by such treatments can be retained to the onset of solidification at rather high cooling rates, thus providing for modification of cast metal.

In many cases a considerable overheating of liquid aluminium melts cannot be achieved under industrial conditions on account of the inadequate power of smelting equipment, low heat resistance of refractory materials, and other technological and economical limitations. As a consequence, the problem arises of how to lower the temperature of homogenization of the melt to acceptable values. This problem can be solved by introducing into the molten alloy small amounts of impurities which reduce the interphase tension at the boundaries of disperse particles of the microheterogeneous melt (this tension determines the thermal stability of particles). Here, as in the case of introduction of cluster-forming additives, we are dealing with the modification of the melt structure, which can improve the quality of cast metal.

Overheating the melt and introducing surface-active additives into it does not exhaust the means of destroying the hereditary microheterogeneity of the melt. Other energy actions on liquid metals – ultrasonic treatment, irradiation by the alternating electromagnetic field or ionizing radiation flows, etc., may also be of practical use. The most acceptable for aluminium alloys, with a comparatively low liquidus temperature, is ultrasonic treatment (UST).

In the following chapters we consider in detail the results of practical implementation of the selected treatments.

Chapter 2

Effect of Heat–Time Treatment (HTT) of Melts on the Structure and Properties of Aluminium Alloys

Simple binary systems, such as Al–Cu, Al–Mg, Al–Si, Al–Zr, Al–Ti, Al–Fe, Al–Cr, Al–Pb, Al–In, and their more complex combinations, constitute the basis for most of the known commercial cast, deformable and granulated alloys.

These systems are characterized by different phase diagrams, among which three major types can be distinguished: eutectics, catatectics, and peritectics. In spite of different chemical compositions of alloys, the basic structural constituents involved in the process of their solidification from the melt are: α-solid solution, eutectic, and crystals of aluminides. It is known that the structure of alloys may be varied over a wide range by changing the solidification conditions. For example, one of the effective methods for improving the structure and raising the properties of materials is high-speed solidification. The great number of experimental studies in this field is testimony of the development of new progressive technologies for producing materials with unique properties [1–3].

Nevertheless, the method of rapid quenching from melts is often limited by purely technical possibilities and is not always practicable for industrial alloy production. In connection with this, the scientific search for additional external actions upon melts (including HTT) has become an important and urgent problem.

In this chapter we present the experimental results of a study of structure-sensitive properties of binary liquid aluminium alloys, in particular, alloys of aluminium with transition metals and silicon, as well as the results of investigations of regularities in the structure formation of melts under HTT conditions and in non-equilibrium solidification.

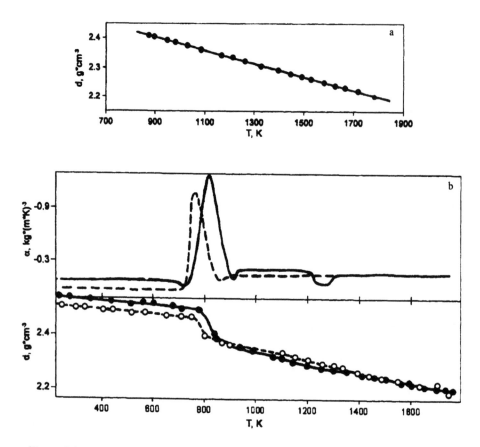

Figure 2.1   (a) Temperature dependences of density obtained in cooling the Al–Si melt of the eutectic composition prepared at 1720 K and (b) temperature dependence of temperature coefficient and density obtained after melting the specimen (•, ○ denote experimental data, lines represent the smoothing spline and its derivative).

2.1   Results of Investigations of Structural Rearrangements in Aluminium-based Liquid Alloys

In the preceding chapter the heat–time treatment of melts was separated out as one of the possible promising ways of upgrading the quality of ingots and castings produced from aluminium alloys. The elaboration of its regimes for particular systems begins with the study of temperature dependences of the properties of the liquid metal and determining characteristic temperatures at which manifestations of structural rearrangements occurring in the metal are detected. This section outlines the main results achieved in this direction.

Systems with simple eutectics: Al–Si, Al–Ge, Al–Sn

A thorough search for the temperatures of structural transformations in *aluminium–silicon* melts, providing a basis for a variety of commercial silumins, was initiated by the densito-

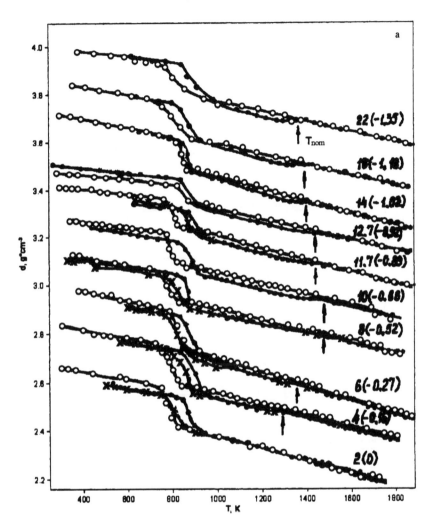

Figure 2.2 (a) Temperature dependences of density (• heating, ○ cooling from 1720 K, x cooling from 970 K) and (b) temperature dependences of thermal expansion coefficient (-·-·-·- cooling from 970 K, ———— cooling from 1720 K, ———— heating) for Al–Si melts. (Digits near the curves show silicon content in at.%, the required shift along the vertical axis is given in parentheses.)

metric investigation [4]. The results obtained are in qualitative agreement with the data on the behaviour of Sn–Pb melts discussed in Chapter 1. When the components of a specimen of eutectic composition were mixed at 1720 K, no specific features that could be associated with the structural rearrangements in the melt were detected on the temperature dependence of density obtained on cooling from this temperature (Fig. 2.1a). However, the densities obtained after solidification and the second melting of the specimen proved to be about 1% lower than before its solidification (Fig. 2.1b). In the temperature range between 1220 and 1380 K the authors observed an anomalous reduction in the thermal expansion coefficient

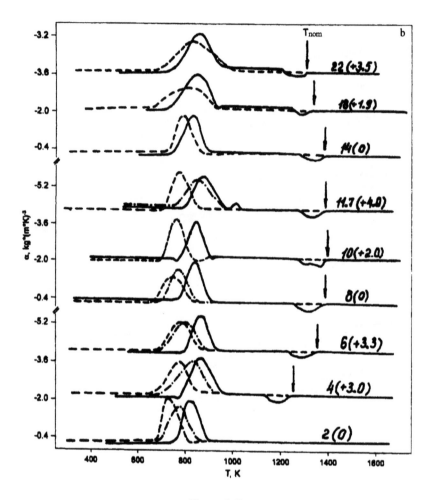

Figure 2.2b

α whereupon the $d(T)$ curve merged with the cooling curve of the initial melt and had subsequently no specific features at any variation of temperature in the range from the eutectic temperature T_e to 1720 K. Such anomalies, but at slightly different temperatures, were also observed for specimens with compositions differing from the eutectic one (Figs. 2.2a, b).

A distinction between the $v(T)$ curves obtained in heating above the melting point to the temperatures in the range 1470–1670 K and subsequent cooling of all specimens except the eutectic one (Fig. 2.3a) was also found in the study of viscosity of Al–Si melts [5]. It is interesting that in the region below the branching point the heating curves for hypo-eutectic compositions run above the cooling curves. In passing through the eutectic concentration, the inversion of viscosity hysteresis is observed: viscosity in cooling exceeds viscosity in heating.

The results obtained were interpreted by the authors on the basis of the concept of the hereditary metastable microheterogeneity of liquid eutectics, while anomalies in the $d(T)$

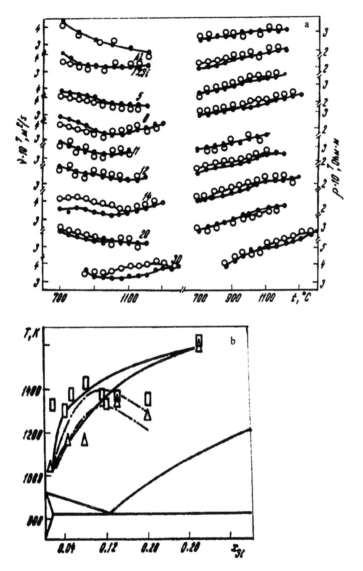

Figure 2.3   (a) Temperature dependences of kinematic viscosity ν and resistivity ρ
(• heating, ○ cooling) and (b) the dome of decay of metastable colloidal microheterogeneity
in Al–Si melts constructed with the use of d (-·-·-), ν (△), and ρ (□) values.

and ν(T) curves were attributed to the irreversible transition of the microheterogeneous
melt, formed upon melting the specimen, into the true solution state. By plotting the points
of density and viscosity anomalies on the phase diagram of the Al–Si system, the
temperature–concentration boundaries of the region of existence of the microheterogeneous states of liquid silumins can be determined (Fig. 2.3b). The lack of branching of
the temperature dependences of viscosity for the specimen of eutectic composition led the
authors to the conclusion that irreversible changes in viscosity are related primarily to the

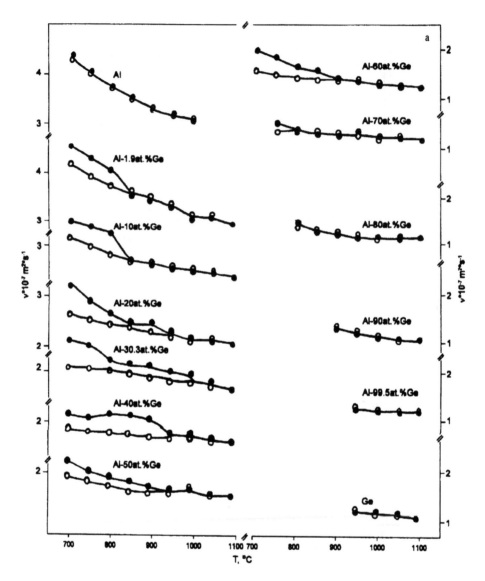

Figure 2.4   (a) Temperature dependences of kinematic viscosity ν (● heating, ○ cooling); (b) the dome of decay of metastable colloidal microheterogeneity; and (c) the $\ln\nu$ versus $1/T$ for Ge melts.

destruction of disperse particles formed from the fragments of the primary phase of the initial ingot. The composition of the primary phase changes in passing through the eutectic concentration, which results in the inversion of viscosity hysteresis.

Grant [3] investigated the temperature dependences of the resistivity ρ of Al–Si melts in heating the specimen upon melting and in the course of subsequent cooling. The results are presented in Fig. 2.3a. For most of the specimens a distinction between the heating and cooling curves was established. The discrepancy begins, as a rule, at the strong overheating

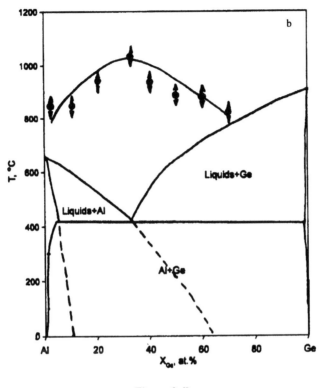

Figure 2.4b

above the liquidus and is observed in comparatively narrow temperature intervals. The branching of the ρ(T) curves was attributed to the preservation of submicroscopic inhomogeneities in the system after the destruction of microheterogeneity of colloidal scale. A slow solution of these inhomogeneities continues in cooling the specimen and at a certain stage of solution their size becomes comparable with the free path length (0.1–1.0 nm [6]), which results in the enhanced scattering of current carriers, accompanied by a decrease in electrical conductivity.

To estimate the volume fraction Φ of the disperse phase in Al–Si melts Korzhavina et al. [5] used Odelevskii's equation relating the electrical conductivity of the two-phase system with conductivities Λ_s and Λ_p of the solvent and disperse particles, respectively:

$$\Lambda = \frac{2\Lambda_s - \Lambda_p + \Phi(3\Lambda_p + \Lambda_s)}{4} + \sqrt{(\frac{2\Lambda_s - \Lambda_p + \Phi(3\Lambda_p + \Lambda_s)}{4})^2 + \frac{\Lambda_s \Lambda_p}{2}}. \quad (2.1)$$

The estimate shows, for example, that for an Al–5 at % Si alloy the Φ value at 970°C is close to 20% and falls to 2% on heating to 1370 K.

The temperatures of structural rearrangements in *aluminium–germanium* melts were determined in [7] by the viscosimetric method. The v(T) dependences are presented in Fig. 2.4a. The discrepancy between the heating and cooling curves observed for all

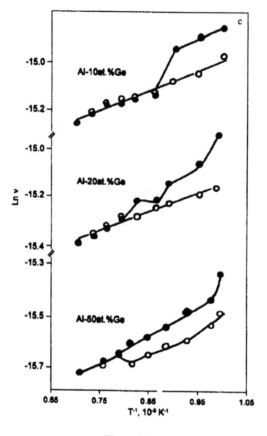

Figure 2.4c

specimens studied in this work attests to the irreversible changes in their structure. The authors related it to the destruction of metastable microheterogeneity arising after melting the alloys, which occurs near the temperatures corresponding to the branching points of the $v(T)$ curves. These points were used to draw the boundary of the region of existence of the microheterogeneous states in Al–Ge melts (Fig. 2.4b).

In order to obtain more detailed information on the specific features of destruction of such states at various Ge concentrations, the authors constructed the $\ln v = f(T^{-1})$ dependences shown in Fig. 2.4c. Based on the results of analysis of these curves, the specimens were divided into three groups with different Ge content: lower than 10 at. %, between 20 and 40 at. %, and above 50 at. %. Alloys of the first group are characterized by a rapid decrease in viscosity in a comparatively narrow temperature range in heating above definite temperatures. On further heating and cooling the specimens, the coincident rectilinear $\ln v = f(T^{-1})$ dependences inherent in regular solutions were revealed. This finding points to the homogeneity of the system. For melts of the second group the viscosity transition to values corresponding to a true solution state occurs gradually over a fairly wide temperature range. In this segment of the curve the derivative $d[\ln v]/d(T^{-1})$ is close to zero, while on further heating and cooling the linear $\ln v = f(T^{-1})$ dependence inherent in regular solutions is obtained. Finally, alloys containing 50 at. % Ge or more display the nonlinear $\ln v = f(T^{-1})$ dependence even upon heating to 1380 K, although the temperature dependences of viscosity obtained in heating and cooling are nearly coincident at $T >$ 1170 K. This is indicative of the essentially non-ideal character of Ge-rich homogeneous solutions. Clearly, in such solutions microheterogeneity of smaller scale is also retained after passing through the dome-shaped curve bounding the region of existence of metastable microheterogenity. The distinction between alloys of the third group and other alloys may be due to the colloid inversion near the equiatomic composition: the dispersive medium turns out to be enriched with germanium, whereas Al-rich disperse particles make up the colloidal suspension in this medium.

The *aluminium–tin* system is a boundary system between eutectic alloys and alloys with miscibility gap. The extended portion of the flat liquidus line on the phase diagram points to a clear-cut tendency for isolation of like atoms.

The temperature dependences of viscosity of these melts obtained in [8] are presented in Fig. 2.5a. The branching of the $v(T)$ curves corresponding to the regimes of heating and subsequent cooling was revealed for the specimens containing from 2.5 to 67 at. % (10–90 wt. %) Sn. This phenomenon is most pronounced in the alloy with 18.5 at. % (50 wt. %) Sn.

The authors related the hysteresis of the temperature dependences of viscosity to the irreversible change in the structure of melts upon heating to definite temperatures. To elucidate the nature of the above rearrangement, the viscosity logarithm was plotted against the inverse temperature and the free energy G_v of activation of the viscous stream was calculated. The results of the calculation for the Al–50 at. % Sn alloy are presented in Fig. 2.6a. They indicate that the linear dependences $\ln v = f(T)$ and $G_v(T)$ inherent in regular solutions are observed only upon heating the melt above the temperature of branching of the $v(T)$ curves. Using the experimentally determined densities of Al–Sn melts, the authors checked the validity of Bachinskii's equation. The linear dependence $v^{-1}(d)$ predicted by this equation was observed only upon heating to the branching temperature (Fig. 2.6b). The corresponding curves obtained on heating the specimens above the melting point have an intricate nonlinear appearance, which attests to the lack of complete mixing of the components on the atomic level.

Analysis of the results obtained by Korzhavina *et al.* [8] leads us to the conclusion that irreversible changes in viscosity of the melts under study are related to their transition to a true solution state. At the same time the initial microheterogeneous state arising after melting can be considered as a metastable emulsion of microscopic droplets enriched with one of the components, in a dispersive medium rich in the other component. The existence of microsegregations of tin in Al–Sn melts is confirmed by the X-ray diffraction measurements [9]. The branching points of the temperature curves of viscosity are used to construct the boundary of the region of metastable microheterogeneity of these melts (Fig. 2.5b). We call attention to the proximity of the concentration boundaries of this region and the flat portion of the liquidus line $T_L(C_{Sn})$: in the region above 70 at. % Sn the derivative $|dT_L/dC_{Sn}|$ rapidly grows with increasing tin content, and the branching of the $v(T)$ curves attesting to microsegregation disappears simultaneously.

To summarize the discussion of the results obtained in the study of simple aluminium-based eutectic systems, we should note that in all the cases considered here the signs of

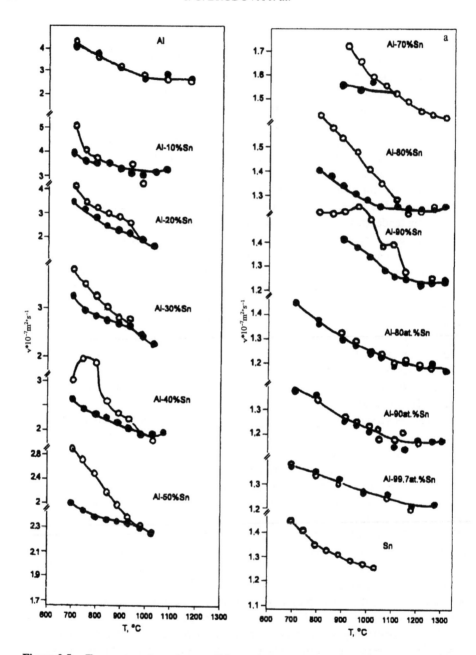

Figure 2.5   Temperature dependences of kinematic viscosity in heating (○) and cooling (●) (a) and the region of existence of the metastable microheterogeneous structure in Al–Sn melts (b). Sn concentrations are given in wt. and at. %.

irreversible destruction of metastable microheterogeneity and transition of the system to the true solution state are observed upon heating to the temperatures specific for each composition. The lack of anomalies in the temperature dependences of properties below

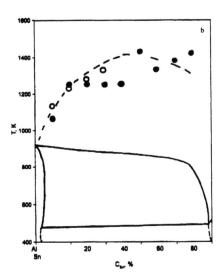

Figure 2.5b

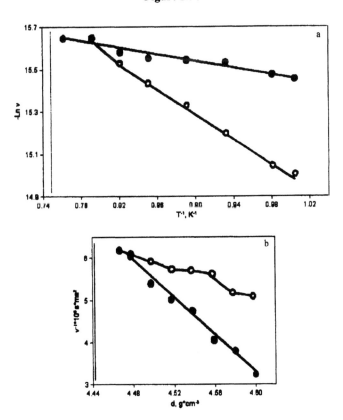

Figure 2.6   lnν as a function of $1/T$ (a) and viscosity ν (b) as a function of free volume for Sn–50 at. % Al melt.

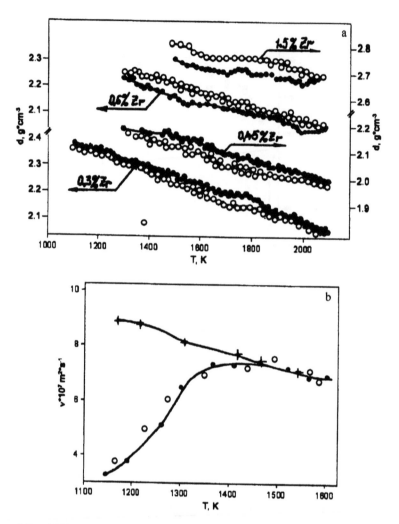

Figure 2.7 (a) Temperature dependences of density of Al–Zr melts (○ heating, • cooling) and (b) temperature dependences of viscosity of Al–0.6 at. % Zr master alloy (• first heating, + cooling, ○ second heating after solidification). Zr concentration is given in at. %.

these temperatures shows that dispersity and composition of colloidal particles varies gradually, without perceptible discontinuities, as the point of the given transition is approached. Therefore, when optimizing the temperature regimes of melting of alloys of this group, the most significant effects should be expected after overheating the system above the temperature of melt homogenization.

Systems with intermetallic compounds: Al–Zr, Al–Ti, Al–Mn, Al–Mg, Al–Sc

The *aluminium–zirconium* system was the first compound-containing system for which the characteristic temperatures of structural rearrangements were determined. Brodova *et al.*

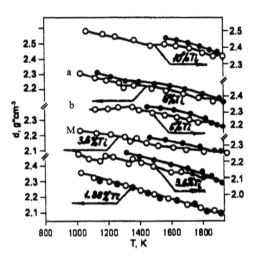

Figure 2.8   Temperature dependences of density of Al–Ti melts. The dependences obtained after the first and second melting of the specimen prepared from Al–6 at. % Ti alloy are marked as •a and °b, respectively. M is the specimen prepared from Al–3.6 at. % Ti alloy solidified under pressure (• heating, ° cooling).

[10] studied the temperature dependences of density and viscosity of the specimens containing up to 1.5 at. % Zr and prepared from high-purity reagents (iodide-purity Zr and Al with an impurity content of no more than 0.001%). The results presented in Fig. 2.7a indicate that in the temperature range from the liquidus to 2070 K, only in the specimen with minimum Zr content, signs of the completion of the system's transition to the true solution state near 1870 K are observed. For higher concentrations of the second component, this transition is not completed by the upper temperature of the range investigated, which is evidenced by the absence of the congruent portion of $d(T)$ curves obtained on heating and subsequent cooling. This conclusion was also confirmed by the viscosimetric study in which alloys of the above purity containing up to 2 at. % Zr were used. In the range of temperatures up to 1820 K no evidence of irreversible changes in the melt structure was found. However, quite non-trivial $v(T)$ dependences pointing to complex processes due to the rise of the specimen temperature were obtained in the study of viscosity of an Al–0.6 at. % Zr master alloy of commercial purity (Fig. 2.7b). In the initial portion of these curves (approximately to 1320 K) the viscosity values increase with heating. Then the growth rate of v slows down and, finally, the "normal" regime of viscosity drop with increasing temperature is established. This high-temperature dependence is retained on further cooling, and below 1470 K the $v(T)$ curve deviates from the curve obtained at the initial heating. Judging by the results presented here, the specimen of commercial purity changes to a true solution state near 1500 K, i.e. at much lower temperature than the point of homogenization of a similar alloy smelted from high-purity components. This fact became a starting point in forming the idea of modifying melts with specially selected additives in order to decrease the temperatures of their homogenization (see Section 3.4).

Figure 2.8 shows the temperature dependences of density of *aluminium–titanium* melts prepared from high-purity materials [11]. The hysteresis of density observed for the

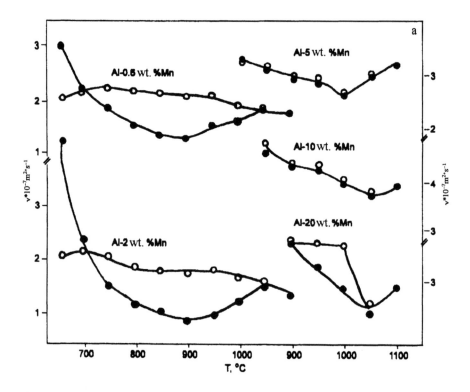

Figure 2.9   (a) Temperature dependences of kinematic viscosity and (b) the dome of the
onset of decay of metastable colloid in Al–Mn melts.

specimens with titanium content exceeding the limiting solubility of Ti in solid Al (about
0.9 at. %) is associated with the destruction of the microsegregated state inherited from the
two-phase crystalline specimen. In this state the melt can be considered as a suspension of
disperse particles with composition close to Al_3Ti in the dilute solution of Ti in Al. The
existence of two such structures of short-range order is supported by X-ray diffraction
studies.

Upon overheating providing the partial or entire solution of disperse particles in the melt,
the initial microheterogeneous state is not recovered, and the solidification process results
in the formation of a structure with finer Al_3Ti precipitates uniformly distributed throughout the volume (see Section 2.2). Upon the second melting of such a specimen, the
branching of the temperature curves of density is less pronounced than in the first experiment (Fig. 2.8). The increase in the titanium content up to 8–10 at. % is also accompanied
by smoothing of hysteresis phenomena, which indicates that in melts of such compositions
higher overheating above the liquidus line is needed for the full solution of the disperse
phase. The size and composition of colloidal particles is related to a certain extent with the
form and the size of the intermetallic compound in the initial alloy, which is evidenced by
the comparison of the $d(T)$ curves of the specimens containing 3.5 at. % Ti, one of which
was melted in a resistance furnace and was solidified at a cooling rate of about 1 K/s,
whereas the second was solidified under the pressure $\sim 10^8$ Pa at a cooling rate of $\sim 10^3$ K/s

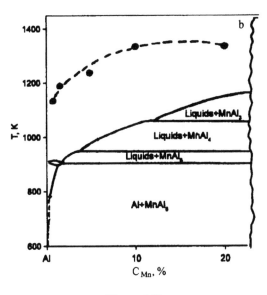

Figure 2.9b

providing a higher, by an order of magnitude, refinement of Al_3Ti precipitates. In the latter case, the branching of the $d(T)$ curves is pronounced less distinctly and the branching point is displaced to the region of lower temperatures (Fig. 2.8).

An important argument in favour of the hereditary origin of microheterogeneity of liquid aluminium alloys is the results of the viscosimetric study of *aluminium–manganese* melts [12]. The phase diagram of this system contains a few compounds. Owing to this, in melting the specimens containing up to 4.1 wt. % Mn, disperse particles with composition close to Al_6Mn are injected into the melt. In the concentration range from 4.1 to 13 wt. % Mn one of the phases of the metastable colloid should have a composition close to Al_4Mn, while at the manganese content from 13 to 40 wt. % Mn its composition is close to Al_3Mn. The authors examined the specimens falling in each of the above concentration ranges.

The results of viscosity measurements are presented in Fig. 2.9a. Analysis of these measurements points to the clear-cut discrepancy of the temperature curves of viscosity obtained in heating and cooling the samples with 0.6, 2.0, and 20 wt. % Mn and to the absence of such a hysteresis for the samples containing 5 and 10 wt. % Mn. The authors related the branching of the $v(T)$ curves to the irreversible transition of the melt from the metastable microheterogeneous state to the true solution state. According to Bibik [13], the viscosity growth on heating, prior to the branching, may be due to the successive dispergation of colloidal particles inheriting the structure of the intermetallic compound. It is apparent that the destruction of microheterogeneity in melts of this system begins at the temperatures corresponding to the minimum in the $v(T)$ curves obtained on heating and ends near the branching point. The points of the minima in the $v(T)$ curves were used to construct the temperature boundaries of the region beyond which the destruction of metastable microheterogeneity begins in Al–Mn melts (Fig. 2.9b).

In the course of experiments on melts of this system, from 10 to 15 measurements of viscosity were made at each temperature. Attention was given to the instability of v values

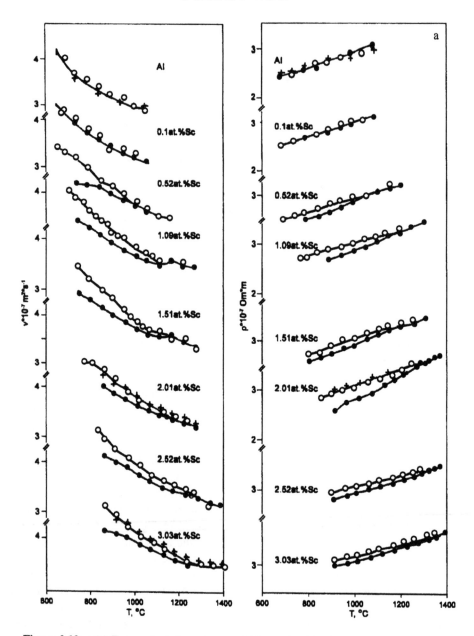

Figure 2.10   (a) Temperature dependences of kinematic viscosity ν and resistivity ρ of Al–Sc melts and (b) temperature dependences of kinematic viscosity ν of the melt obtained with the use of Al–0.6 at. % Sc master alloy (● heating, ○ cooling).

during the few minutes after thermal equilibrium was established. Since this phenomenon was observed only in the regime of heating up to the temperatures corresponding to the minima in the ν(*T*) curves, the authors ascribed it to the change of dispersity and composition of colloidal particles with time.

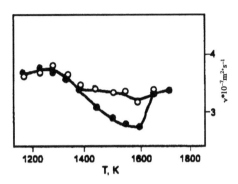

Figure 2.10b

The results of viscosity measurements in [12] were used to construct the $\ln v = f(T^{-1})$ dependences and the temperature curves $G_v(T)$ of the free energy of activation of the viscous flow for the specimens containing 0.6 and 2 wt. % Mn. On heating above the melting point, these activation energies determined from the slope of the $\ln v = f(T^{-1})$ curves are greater by a factor of 10 to 13 than the energies obtained on subsequent cooling. Overheating melts above the points where the branching of the temperature curves of viscosity occurs, changes the sign of the derivative dG_v/dT. These facts comply with the radical change in the pattern of the viscous flow. This agrees with the concept developed by the authors of the work [12] that in the temperature range from the liquidus to the branching point of $v(T)$ curves the disperse particles with composition close to the Al_6Mn compound manifest themselves as structural units of the flow, whereas in cooling the homogeneous solution these structural units are presented by individual atoms of both components.

The absence of hysteresis of the temperature curves of viscosity at 5 and 10 wt. % Mn indicates, in the authors' opinion, that the destruction of colloidal disperse particles of the melt inheriting the structure of Al_4Mn occurs over a wider temperature range than the destruction of particles whose composition is close to Al_6Mn and Al_3Mn. According to [14], the stability of compounds is determined by the filling of the d shell of a transition metal which is at a maximum in the Al_4Mn compound.

The temperatures of structural rearrangements in *aluminium–scandium* melts were determined in [15]. The authors studied the temperature dependences of viscosity and electrical resistivity of specimens containing up to 3 at. % Sc. The results presented in Fig. 2.10a give evidence of irreversible changes of the structural state of Al–Sc melts upon heating to the temperatures specific for each composition. These rearrangements are interpreted as the transition of the system from the microheterogeneous state to the true solution state. Disperse particles are likely to have a composition close to Al_3Sc, while the dispersive medium represents a dilute solution of scandium in liquid aluminium.

Microalloying of aluminium alloys with high-melting elements is implemented in industrial conditions by the introduction of preliminarily melted, concentrated master alloys in the alloys.

It was of interest to compare the influence on the temperature of melt homogenization of equal amounts of impurity elements introduced into the alloy in the form of pure metals and master alloys. This is possible if we compare the viscosimetric experiment described

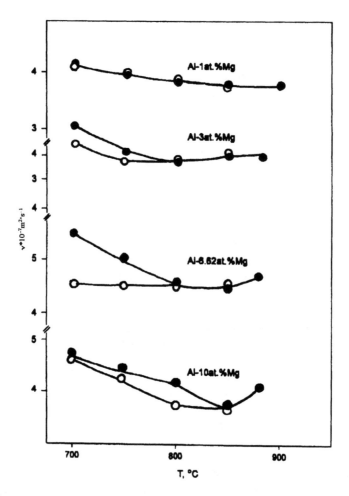

Figure 2.11   Temperature dependences of kinematic viscosity of Al–Mg melts (• heating, ○ cooling).

above, in which Al–0.6 at. % Sc alloys were prepared by smelting of pure aluminium and scandium, with the data [16] obtained by measuring the viscosity of the liquid master alloy.

From the results presented in Figs. 2.10a and 2.10b it follows that the homogenization temperature of the master alloy is higher by 150 K. This effect is accounted for by the fact that the concentration of the refractory element in the master alloy exceeds greatly its limiting solubility in solid aluminium, and by the presence of high-melting aluminides Al_3Sc in the structure. It is evident that upon solution of the master alloy in the matrix system the particles of the intermetallic compound, which exhibit high thermal stability, come into equilibrium with the surrounding melt after a series of dispergation events. Microheterogeneity formed on the basis of such particles is destroyed at higher temperatures than the heterogeneity formed by Sc-rich particles in the dispersive medium of aluminium.

The important regularities observed enabled the authors to explain a series of non-trivial results concerning the structure formation in alloys of aluminium with transition metals.

The compositions of liquid aluminium compound-containing alloys considered above were referred mainly to compositions exceeding the limiting solubility of the high-melting element in aluminium. In a similar way, the behaviour of melts in heating and cooling was studied in the case when the alloys represented aluminium-based solid solutions.

For example, the viscosimetric method was used to determine the temperatures of structural rearrangements in *aluminium–magnesium* melts containing up to 10 at. % of the second component. The temperature dependences of viscosity are presented in Fig. 2.11. For the specimens containing three or more atomic per cent of magnesium, the observed branching of the heating and cooling curves is related, in the authors' opinion, to the irreversible transition of the microheterogeneous melt to the true solution state. It is likely that in this case the metastable colloid is formed by disperse particles of molten Al_3Mg_2 compound, which are suspended in a dilute solution of magnesium in aluminium.

For the specimen containing 6.62 at. % Mg the Odelevskii's equation (2.1) was used to estimate the volume fraction Φ of disperse particles. According to this estimate, Φ decreases from 2.6% to 0.7% as the temperature rises from 970 to 1070 K, while in subsequent cooling Φ is close to zero. Moreover, the linear size of the cell of the melt per one disperse particle was estimated on the basis of the Frenkel–Eiring model [10]: at 970 K it proved to be close to 200 nm.

The results of experiments and calculations performed in this study show that even at concentrations not exceeding the limiting solubility of magnesium in solid aluminium, a microheterogeneous melt forms on melting. Although the temperatures of melting of compounds in this system lie below the melting points of pure components, the melt changes to a true solution state only after heating above the liquidus by 100–200 K.

Thus, in the study of the temperature dependences of properties of liquid aluminium compound-containing alloys, along with the features of irreversible transitions from the metastable microheterogeneous state to the state of true solution, additional anomalies are detected in some cases in the heating curves. These anomalies indicate the possibility of structural transformations in the microheterogeneous system, such as abrupt changes in size, composition, and the interior structure of colloidal particles.

Systems with miscibility gap: Al–In, Al–Pb

As was pointed out in Section 1.6, indications of the metastable microheterogeneity are also found in systems with a miscibility gap above the dome of their macroscopic separation. Popel *et al.* [17] studied the conditions for destruction of such states in *aluminium–indium* melts. The signs of existence of colloidal-scale microheterogeneity in these melts at small overheating above the dome were noted in electron diffraction and ultraacoustic experiments [18], as well as in the study of small-angle X-ray scattering [19].

When measuring the viscosity of these melts in the course of primary heating, the authors revealed an anomalously high scatter of its values (to within 10–15%) which was retained up to temperatures specific for each composition. On further heating and subsequent cooling stable viscosity values were observed up to the boundary of the region of macroscopic separation. In the authors' opinion, the observed instability of the v values is related to the fact that in going beyond the limits of the immiscibility dome the system changes from the macroscopically heterogeneous state to the metastable microheterogeneous state

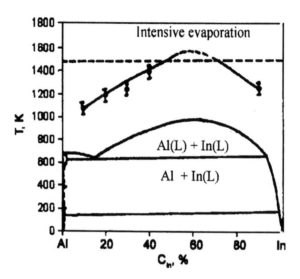

Figure 2.12   Part of the phase diagram and the dome of decay of metastable colloidal microheterogeneity of the Al–In system.

of the microemulsion type, which in turn is destroyed on heating to the temperatures indicated above.

By plotting the temperatures at which viscosity values are stabilized on the phase diagram of the Al–In system, the authors obtained a dome-shaped curve bounding the region of existence of the metastable colloidal structure (Fig. 2.12).

The long-term instability of viscosity of melts in this system was also observed by Herwig and Hoyer [20]. These investigators have managed to stabilize the results of measurements after prolonged isothermal holding of the melt. It was found that at a temperature 200 K above the immiscibility dome viscosity is stabilized only after a lapse of three days. These data seem to be the first experimental estimate of the lifetime of the metastable microheterogeneous state of the melt far away from the boundary of the region of its existence and give evidence of the rather high stability of the metastable colloid.

More recently, Chikova *et al.* [21] performed similar studies for melts of the *aluminium–lead* system which constitute the basis of a series of commercial antifriction alloys. Strong evaporation of the second component did not allow the authors to carry out experiments with specimens containing more than 10 at. % Pb. At such concentrations, the instability of viscosity values in primary heating is less pronounced than in alloys with indium. However, the authors revealed branching of the $v(T)$ curves obtained in heating and subsequent cooling, which indicates that in this case the melt also changes irreversibly from the metastable microheterogeneous state inherited from the macroscopically separated initial system to the true solution state (Fig. 2.13).

The above data on the temperatures of structural rearrangements in aluminium-based melts were used to choose the regimes of HTT of melts for producing ingots and castings. The results of these studies are presented in the following sections of Chapter 2 and in Chapter 3.

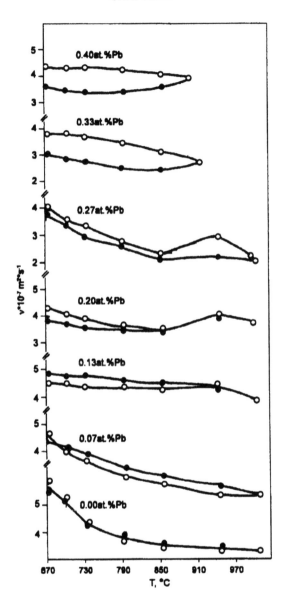

Figure 2.13   Temperature dependences of kinematic viscosity of Al–Pb melts (• heating, ○ cooling).

2.2   Some Regularities of the Structure Formation in Alloys Obtained by Solidification from Microheterogeneous Melts

As follows from the results presented in Section 2.1, aluminium melts retain the micro-heterogeneous state over a wide range of temperatures above the liquidus T_L, and the most significant structural rearrangements in them are observed only on heating above T_{hom}. For this reason, it makes sense to consider the specific features of formation of the solid phase under two melt processing temperature regimes, namely, when the temperature of heating the melt does not exceed the homogenization temperature ($T_L < T < T_{hom}$) and when the temperature of heating is in excess of the homogenization temperature ($T > T_{hom}$). This conditional division is fundamentally important because it provides insight into the physical basis for interrelation between the liquid and solid metallic states.

To date ample experimental material has been accumulated, suggesting that nucleation of the solid phase is related to the existence in melts of inhomogeneities of different kinds [22, 23]. In this case a change in the state of the melt with an increase in the heating temperature and variation in the chemical composition should result in changes in the distribution of crystallization centres, their size, and growth rate. The idea of such an approach calls for further refinement of the theory of formation of the solid phase from the melt, because its basic assumptions rest on the classical concept about the melt as a micro-heterogeneous system.

If we develop this line of reasoning, it will be clear that in real melts the processes of initial nucleation and the growth of nuclei to the critical size are significantly facilitated, which leads to changes in the equilibrium concentration of potential crystallization centres and in the critical overheating required for a significant rate of heterogeneous nucleation.

Such an approach to the description of the nucleation process also introduces the new elements in our understanding of the processes of atomic exchange at the boundary of the phase transition, which provide the growth of the crystal from the melt.

The physical foundations of the process of crystal growth are discussed in detail in the fundamental works of Cahn [24], Kramer and Tiller [25], Chalmers [26], Jackson [27], and Mullins [28]. The essence of these works is that the microtopology of the interface between the liquid and solid phases determines the mechanisms of the growth of the latter, which in turn determine the particular habitus of crystals and the kinetics of their growth. A distinct interrelation exists, in addition, between the solidification rate and the degree of deviation of the system from equilibrium, i.e. the value of the melt supercooling.

These arguments concern primarily the solidification of pure metals. The description of the phase transition in alloys is significantly complicated because of the change of equilibrium conditions at the interphase boundary. Also, along with thermal supercooling of the melt, an additional source of driving force for the crystallization process appears, namely, the concentration supercooling related to the interaction between the components in the system and to their phase diagram [26, 29].

However, the basic postulates of the theory of crystal nucleation and growth are common for both cases. According to the most commonly used Jackson's criterion for estimating the mechanisms of crystal growth, crystals of α-solid solution and crystals of aluminides of transition metals represent substances with different entropy of fusion and therefore, even under the same solidification conditions, their growth is governed by different laws. The last circumstance imposes constraints on the external shape, sizes, and the number of phases present in the alloy structure. Since the laws governing the formation of α-dendrites, aluminides, and eutectic are different, it is useful to separate them and to analyze the effect of HTT of melts and solidification parameters on each structural constituent separately. A knowledge of these features of structure formation will enable one to effectively control the structure and properties of castings and ingots.

The specimens were obtained by quenching from the liquid state. It is known that quenching is referred to as a thermal operation by which a metal or alloy is heated above the transformation temperature with subsequent, rather fast cooling. As a result of rapid cooling, the high-temperature equilibrium state of the alloy or its intermediate metastable state is fixed. In contrast to ordinary quenching from the solid state, the term "quenching from the liquid state" denotes the technological operation related to the rapid cooling of the melt. The main cooling method in this case is heat removal through the solid support. A decrease in the thickness of the liquid film and an increase in the heat transfer coefficient at the boundary between the film and the heat-conductive support make it possible to obtain the cooling rates as high as 10^{10} K/s.

Brief description of methods for producing specimens

We used general procedures of quenching from the liquid state providing cooling rates in the range $10^2 - 10^6$ K/s. The specimens in the form of ribbons $100-500$ μm thick were obtained by the "splat cooling" technique, i.e. by spreading out the liquid alloy over the internal surface of a rotating copper cone and by its subsequent cooling and solidification. In this method the one-sided cooling of the melt through the heat-conducting support is employed. The ribbon thickness was set by the rate of cone rotation; the specimen cooling rate was estimated by a calculation procedure using the known ribbon thickness [30].

The specimens in the form of disks (films), with variable diameter and thickness depending on the cooling rate, were obtained by "splattering" a droplet of small mass between two massive copper tips. The specimen was placed on a mica support and was moved to the heating or cooling zone with the aid of special rods. Heating of the melt was accomplished in a vacuum using a molybdenum heater with water-cooled current leads.

The direct control and correct estimate of the rate of melt cooling were made using a specially devised method of recording the cooling curve on the display of an S8-13 memory oscillograph. The temperature was measured by a tungsten–rhenium thermocouple mounted in the heated zone of the melt.

All the above technical measures allowed us to estimate separately the role of various solidification parameters in the formation of the structure of aluminium alloys. For example, the effect of the melt overheating ΔT was investigated on specimens solidified at the fixed cooling rate, whereas the role of the cooling rate in forming the structure was studied at constant overheating of the melt above T_L.

The samples obtained at low cooling rates ($V = 1 - 100$ K/s) were produced by casting the melt into graphite, alundum, or metallic ingot moulds. Fusion and heat–time treatment of the melt were made in a chamber furnace with silit heater and thermoregulator.

Thus, the experimental procedures described above permitted us to provide high heating (up to 1600 K), temperature control, and solidification of liquid aluminium alloys in a wide range of cooling rates from 1 to 10^6 K/s.

2.2.1. Morphology and kinetics of growth of silicon crystals

The changes of morphology, size and quantity of primary silicon depending on the solidification conditions (cooling rate and melt overheating) were studied on the alloy of

Figure 2.14   Growth forms of silicon crystals in Al–26% Si alloy as a function of melt overheating: (a) $\Delta T = 120$ K; (b) 420 K ($\times 2000$).

hypereutectic Al–26% Si composition* ($T_L = 1020$ K). According to the equilibrium phase diagram, such a silicon content in the alloy provides a high volume fraction of primary silicon crystals.

As follows from the data presented in Section 2.1, the melts of hypereutectic silumins are characterized by a clear-cut microheterogeneous structure, which is retained over a wide temperature range of overheating above T_L. In connection with this, the temperature regime of treatment of the liquid Al–26% Si alloy varied from 1070 to 1670 K, which corresponded to $\Delta T = 50$–650 K, while the cooling rate in the solidification of disk-shaped specimens was 10^2–10^4 K/s.

The shape and dimensions of primary silicon crystals were determined by the method of quantitative metallography, the results of which are presented in Figs. 2.14a,b and 2.15a. It was established that when the temperature of heating the melt was in excess of 1470 K, we

---

* Here, and henceforth in Chapters 2 and 3 concentrations are given in weight per cent.

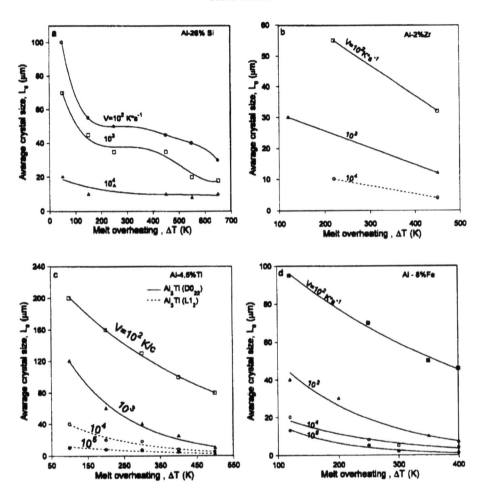

Figure 2.15 Dependence of the average size of silicon crystals (a) and aluminides of zirconium (b), titanium (c), and iron (d) on solidification conditions.

observed a change in morphology of crystals of primary silicon. This fact is clearly seen on the microphotographs of the cast structure of specimens obtained with $V = 10^3$ K/s (Figs. 2.14a,b). Faceted crystals of irregular geometric shape are replaced by branched dendritic crystals. An increase in the cooling rate also stabilizes dendritic forms of the growth, whereas the effect of overheating reduces in this case to a decrease in the size and number of primary crystals. A similar reduction of the linear size of crystals with increasing ΔT was obtained for all rapidly cooled specimens; the tendency being stronger for crystals obtained with lower cooling rates (Fig. 2.15a).

To verify these trends in the growth of silicon crystals a hypereutectic alloy with lower silicon content (17%) was studied in [31]. In spite of the decrease in the silicon content, the experimental data characterizing the structural state of such a melt show with certainty that the melt is microheterogeneous and, therefore, prerequisites exist for controlling the structure by means of HTT of liquid silumin.

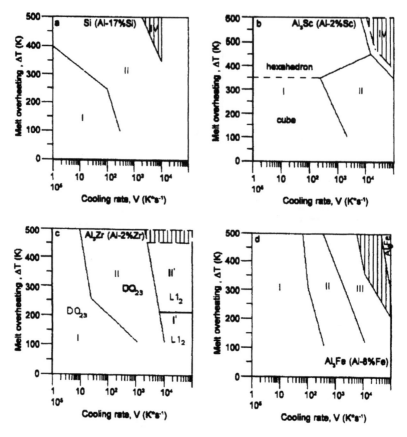

Figure 2.16   Zones of growth of crystals with different morphology: (a) Si; (b) scandium aluminide; (c) zirconium aluminides (I, faceted crystals; II, dendritic crystals; III, spherulite-like crystals; IV, globular crystals).

The specimens of the Al–17% Si alloy were obtained with the use of various technologies: by casting into a graphite mould ($V < 100$ K/s) and by quenching from the liquid state ($V = 10^2$–10^4 K/s). The range of overheating temperatures was 100–500 K and all the temperatures were below T_{hom}.

As follows from Fig. 2.16a, the dendritic growth of crystals is most stable in wide ranges of V and ΔT, whereas the transition to faceted crystals, growing as polyhedra of irregular shape, is observed only in the range of low temperatures of the melt heating and at $V < 100$ K/s (region I). High-temperature treatment of the melt ($\Delta T > 350$ K) facilitates the phase dispergation and formation of more equiaxed, nearly globular crystals ($V > 10^3$ K/s, region IV). The effect of transition from faceted crystals to dendritic and globular forms with increasing ΔT, which was revealed in this study, disappeared if the melt (overheated in advance) was precooled to a temperature slightly above T_L.

Thus, the decrease in the silicon content to 17% changes the number, size, and morphology of crystals of the primary phase, with retention, however, of the previously determined characteristics of their transformation depending on solidification conditions.

2.2.2. Morphology and kinetics of growth of aluminide crystals

Crystals of primary intermetallic compounds – aluminides – are important structural components of alloys of aluminium with metals exhibiting a limited solubility in aluminium, such as Cr, Ti, Mn, Fe, Sc. Their formation results in the creation of a heterophase structure, which is undesirable for a large group of alloys because of the reduction in the degree of alloying of the α-solid solution, and therefore in its strength. Moreover, brittleness increases and plasticity characteristics are lowered.

However, we can single out another group of alloys – master alloys, for which the morphological features of aluminide crystals is an important factor determining the functional capability of such alloys.

As follows from Table 2.1, the experimentally selected chemical compositions of binary model alloys enable one to describe the diversity of structures and characteristics of the growth kinetics for the most typical intermetallic crystals formed in commercial granulated and casting aluminium alloys.

Table 2.1 Classification of aluminides by crystal structures [32].

Compound	Structure type	Space group	Lattice type	Alloy composition
Al_6Mn	Al_6Mn	C_{mcm}	orthorhombic	Al–5%Mn, Al–2%Mn–2%CrC
Al_3Fe	Al_3Fe	F_{mmm}	orthorhombic	Al–8%Fe
Al_3Ti	Al_3Ti, DO_{22}	I 4/mmm	tetragonal	Al–4.5%Ti
Al_3Zr	Al_3Zr, DO_{23}	I 4/mmm	tetragonal	Al–0.6%Zn, Al–1.5%Zr, Al–2%Zr, Al–3%Zr
Al_3Sc	Cu_3Au, Ll_2	Pm3m	cubic primitive	Al–2%Sc
Si	diamond A_4	Fd3m	cubic fcc	Al–17%Sc, Al–26%Si
Al_7Cr			monoclinic	Al–0.6%Cr, Al–5%Cr

The results of investigations of the mechanisms and kinetics of growth of aluminide crystals under non-equilibrium solidification conditions were obtained in the study of the structure of hypereutectic (Al–Sc, Al–Fe, Al–Mn, Al–Si) and hyperperitectic (Al–Zr, Al–Ti, Al–Cr–Zr) alloys by the methods of optical and electron microscopy, X-ray diffraction, and X-ray spectral analysis [10, 12, 16, 33–35].

We consider below the experimental data in greater detail.

Crystals with cubic lattice

The intermetallic compound Al_3Sc, which is the primary phase in solidification of a hypereutectic Al–2% Sc alloy, was chosen as a typical representative of aluminides with a primitive cubic lattice (structure type Ll_2). The specimens were obtained by quenching from the liquid state according to the procedures described above in the range of cooling

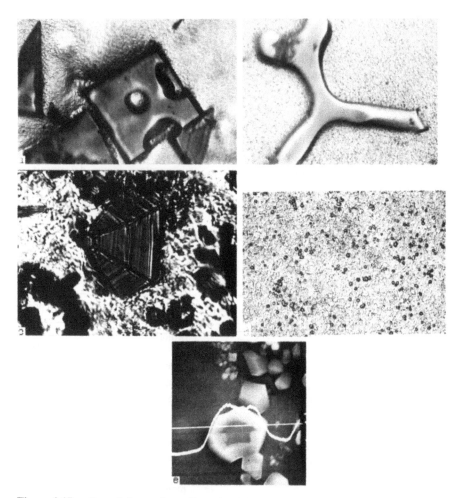

Figure 2.17   Growth form of scandium aluminide crystals as a function of solidification parameters of Al–2% Sc alloy: (a) $V = 10$ K/s, $\Delta T = 100$ K; (b) $V = 10^3$ K/s, $\Delta T = 300$ K; (c) $V = 10^5$ K/s, $\Delta T = 300$ K; (d) $V = 10$ K/s, $\Delta T = 350$ K; (e) secondary electron image with scandium distribution line ($\times 1200$).

rates from 10^2 to 10^5 K/s and at temperatures between 1170 and 1670 K, i.e. with $\Delta T = 100$–600 K. According to the data presented in Section 2.1, this interval of ΔT covered the entire range of existence of the metastable microheterogeneous state of the liquid alloy.

The results of X-ray diffraction analysis showed that for all solidification parameters under investigation aluminides formed in the alloy structure have an Al_3Sc composition and a crystal lattice of the $L1_2$ structure type with the lattice parameter $a = 0.4106$ nm.

Metallographic studies revealed the following changes in the morphology of crystal growth. When the melt is overheated to $\Delta T < 300$ K, primary aluminides in castings obtained with $V \leq 100$ K/s retain the polyhedral growth form, mainly the cubic form, and have a size of 30–50 μm (Fig. 2.17a). An increase in V to 10^3–10^4 K/s leads to the transformation of faceted growth forms, which are replaced by dendritic forms; the degree of branching of dendrites grows with increasing V (Fig. 2.17b). At $V = 10^5$ K/s, the dendritic growth is suppressed, crystals become smaller, reaching 5–7 μm in size and take on a nearly spherical shape (Fig. 2.17d). Therefore, an increase in the cooling rate results in the following succession of crystal growth forms: faceted crystals of cubic shape, dendrites, spherical crystals.

Preliminary heating of the melt to 1370–1570 K ($\Delta T > 300$ K), under the same cooling and solidification conditions, causes a drastic change in the morphology of primary aluminides. In this case Al$_3$Sc crystals are found to grow in the form of regular hexahedral truncated pyramids (Fig. 2.17c). This inference is also supported by the results of EPMA shown in Fig. 2.17e. According to theoretical and experimental studies of the anisotropy of mobility of the crystal–melt interphase boundary [42], the transformation of the crystal growth form from a cube to a hexahedral pyramid is observed in connection with the change of the $\langle 100 \rangle$ crystallographic orientation of the preferential crystal growth to the $\langle 111 \rangle$ orientation.

As the cooling rate V increases, crystals become smaller, but retain specific features of the shape formation: for example, at $V = 10^4$ K/s they have a size of 5 μm and a nearly spherical shape. Further increase in ΔT and V gives rise to a change in the kinetics of crystal growth, which manifests itself in reducing their dimensions to 1–3 μm. The sequence of regions of stable growth of Al$_3$Sc crystals, depending on V and ΔT, is shown schematically in Fig. 2.16b; we see that the variation in the conditions of solidification of aluminides with cubic lattice allows one to transform their growth forms and sizes.

Crystals with tetragonal lattice

We studied the alloys of hyperperitectic compositions containing 0.6, 1.5, 2.0, 3.0, and 4.7% Zr. The specimens were obtained in the form of castings at low V and in the form of ribbons at high V. Let us consider first the peculiarities of the structure formation of such compositions at low overheating of the melt above the liquidus temperature, i.e. at $\Delta T \leq 150$ K.

At low cooling rates intermetallic compounds of the Al$_3$Zr composition, having, according to the equilibrium phase diagram, a tetragonal crystal lattice (DO_{23} space group) form in castings independently of the concentration of the second component. The calculated lattice parameters $a = 0.4013$ nm and $c = 1.732$ nm agree well with the available data [32].

The most typical form of crystal growth is faceted elongated plates, the size of which decreases with the growth of the cooling rate (Fig. 2.18a). As the Zr concentration in the alloy increases, the distribution of crystals over the specimen cross-section becomes more and more non-uniform and the mean size increases from 60 to 250 μm. Intermetallic compounds of this modification grow steadily in a definite range of cooling rates, which depends on the alloy composition. For an Al–1.5% Zr alloy this range corresponds to 10^{-1}–10^2 K/s; for Al–2% Zr, to 10^{-1}–10^3 K/s; for Al–3% Zr, to 10^{-1}–10^4 K/s, and for Al–4.7% Zr, to 10^{-1}–10^5 K/s.

It was of interest to study the influence of the state of the melt on the morphological stability of growth and the sizes of such crystals. With this purpose, the Al–2% Zr melt was overheated to various temperatures ($\Delta T = 100$–450 K). The size of stable zirconium

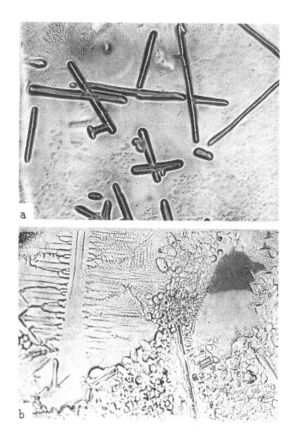

Figure 2.18   Growth form of zirconium aluminide crystals (the structure type $D0_{23}$) as a function of solidification parameters for Al–2% Zr alloy: (a) $V = 10$ K/s, $\Delta T = 200$ K; (b) $V = 10^3$ K/s, $\Delta T = 360$ K ($\times 500$).

aluminides is plotted against V and T in Fig. 2.15b and the change in their shape is shown in Fig. 2.16c (regions I and II). Figure 2.18b illustrates a typical case, in which dendritic crystals with distinctly outlined facet vertices of the first-order and second-order branches are formed in place of plates. Attention should be paid to the formation of much more disperse equiaxed dendrites having a specific "petal-shaped" structure; their morphology and sizes give unambiguous evidence of their primary origin.

Let us consider specific features of their formation in greater detail. The results of our investigations show that for each composition of binary Al–Zr alloys (0.6, 1.5, 2.0, 3.0, 4.7% Zr) a particular range of cooling rates exists, in which primary intermetallics of similar morphology, structure and sizes form.

X-ray phase analysis of such crystals showed that they have the Al_3Zr composition, and a cubic ordered structure similar to those for the secondary metastable phase precipitated in the process of decomposition of the oversaturated α-solid solution [36–38].

As the Zr concentration in the alloy increases, the conditions for the most stable growth of intermetallic compounds of the metastable modification shift toward higher cooling rates (Table 2.2).

Table 2.2   Conditions for the formation of Al$_3$Zr aluminides of the metastable modification.

Zr concentration, %	Minimum cooling rate, K/s
0.6	10^2
1.5	10^3
2.0	10^4
3.0	10^5

If cooling rates are lower than those indicated in this table, intermetallics of stable modification grow in the structure of alloys of the given compositions.

In our experiments [33, 35] the effect of HTT of the melt on the morphological stability of the forms of growth of metastable aluminides was first examined. We studied the influence of heating the melt on the size, morphology, and structure of aluminides in an Al–2% Zr alloy. The melt was overheated above the liquidus temperature by 160–460 K.

It was found that at $\Delta T = 100$ K aluminides grow in the form of faceted crystals of cubic shape (Fig. 2.19a, region I′ in Fig. 2.16c). At high ΔT the dendritic form becomes the dominant form of crystal growth (region II′ in Fig. 2.16c). At $\Delta T = 200–250$ K intermetallic compounds have a maximum size as large as 10 μm and grow as dendrites with strongly pronounced anisotropy of the rate of growth of primary and secondary branches (Fig. 2.19b). At $\Delta T = 400$ K, the size of crystals reduces to 5 μm and they take the shape of symmetric dendrites (Fig. 2.19c). Heating to the temperature region T_{hom} is accompanied by a sharp increase in the amount of aluminides and an additional decrease in their size to 1–2 μm (the hatched area in Fig. 2.16c).

Thus, by varying the melt preparation conditions we can change the size (Fig. 2.15b) and morphology of crystals of the metastable Al$_3$Zr phase over a wide range.

Taking into account the extensive use of Al–Ti alloys as master alloys, it was interesting to investigate in detail their structure formation under non-equilibrium solidification conditions.

According to the basic phase diagram of the Al–Ti system, an intermetallic compound of Al$_3$Ti composition with a tetragonal lattice of the $D0_{22}$ structure type forms peritectically in the range of Al-rich compositions.

The alloys of Al–4.5% Ti and Al–6% Ti compositions were used as objects for the study. In preparation of the specimens by the methods described above, the solidification conditions were varied in the following way: ΔT in the range from 100 to 500 K and V in the range $10–10^5$ K/s.

X-ray diffraction analysis in combination with metallographic examination showed that Al$_3$Ti crystals with a plate structure, similar to the Al$_3$Zr crystals considered above, form in the alloys of both compositions in the range of cooling rates from 10 to 10^3 K/s.

Overheating of the melt does not change the form of crystal growth; only a decrease in the crystal size is observed. The dependence of the size of intermetallics in the structure of the Al–4.5% Ti alloy specimens obtained at various cooling rates on heat treatment of the melt is shown in Fig. 2.15c. It is evident that the appearance of the curve is identical for all specimens: an increase in the overheating favours dispergation of primary phases but the most dramatic change in size is observed when solidification occurs at cooling rates $V > 10^3$ K/s. In this case the morphology of aluminide crystals changes. At $V = 10^3$ K/s they

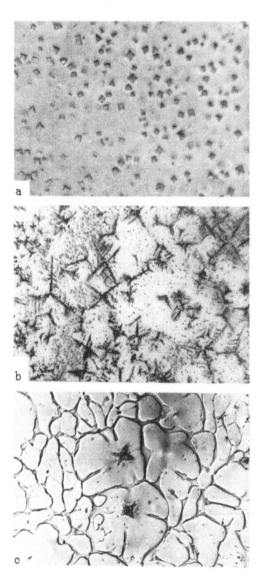

Figure 2.19 Transformation of the forms of growth of zirconium aluminide crystals (the structure type $L1_2$) as a function of melt overheating ($V = 10^4$ K/s): (a) $\Delta T = 100$ K; (b) $\Delta T = 300$ K ($\times 800$); (c) $\Delta T = 450$ K ($\times 2000$).

grow in the form of short needles, the size of which continues to decrease with increasing temperature and reaches 20 μm (Fig. 2.20a). An increase in the cooling rate to $V > 10^3$ K/s leads to the formation of dendritic crystals of intermetallic compounds. At low overheating of the melt anisotropic dendrites of a stable Al_3Ti phase grow, while an increase in the temperature to $T > 1370$ K ($\Delta T = 300$ K) causes a change not only in the crystal morphology but also in their crystal structure (Fig. 2.20b). According to selected area diffraction patterns (SADPs), the crystals have a cubic lattice of the $L1_2$ type with the lattice parameter

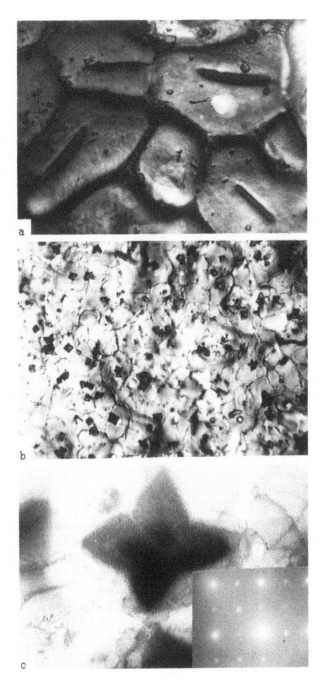

Figure 2.20   Growth form of titanium aluminide crystals as a function of solidification parameters for Al–4.5% Ti alloy: (a) a stable Al_3Ti phase with DO_{22} tetragonal lattice, $V = 10^3$ K/s, $\Delta T = 420$ K ($\times 1200$); (b) a metastable Al_3Ti phase with $L1_2$ cubic lattice, $V = 10^5$ K/s, $\Delta T = 120$ K ($\times 1200$); (c) the same phase, electron microscopy, bright-field image and selected area diffraction pattern ($\times 15\,000$).

$a = 0.4041$ nm. The forms of crystal growth of this phase are fairly different and vary with solidification conditions much like the metastable zirconium aluminides described above. Preliminary heating of the melt into the region T_{hom} ($\Delta T = 400$ K) and its subsequent solidification with the rate $V = 10^5$ K/s determine the kinetics of nucleation and growth of crystallization centres which provide the formation of the disperse structure containing metastable Ti aluminides of $1-2$ μm size with a cubic lattice.

An increase in the Ti content in the alloy up to 6% does not change the character of its solidification. However, all regions of morphological stability of the growth of primary phases considered above for the Al–4.5% Ti alloy shift to higher V and ΔT. For example, the formation of the metastable Al_3Ti phase is observed only at $V = 10^5$ K/s, whereas the region of its existence extends to very high temperatures including the maximum temperatures possible under the given experimental conditions.

Crystals with orthorhombic lattice

Manganese and iron aluminides Al_6Mn and Al_3Fe (Table 2.1) were selected to investigate morphology and kinetics of the growth of intermetallic crystals with orthorhombic lattice. The equilibrium phase diagrams of the Al–Fe and Al–Mn systems in the Al-rich regions are known to be identical. The alloys solidify as eutectic alloys under equilibrium conditions; the eutectic points correspond to insignificant amounts of the second components (about 2%); the solubilities of Mn and Fe in the aluminium-based α-solid solution are small. All these factors permitted us to join the results concerning the structure formation in hypereutectic alloys of these systems, whose phase constitution includes a eutectic and crystals of primary aluminides.

We checked the effect of overheating on the morphology of the crystal growth in alloys obtained at low cooling rates ($V = 1$ K/s). For this purpose we investigated the structure of an ingot prepared from the Al–5% Mn alloy and established that at low temperatures of heating the melt above the liquidus ($\Delta T = 120$ K) the primary intermetallics grow as polyhedra of irregular shape. Overheating the liquid metal ($\Delta T = 320$ K) causes a change in the morphology of growing crystals, and hexahedral growth forms are observed in the alloy structure. Such a transition points to the regular change of direction of the preferential crystal growth from $\langle 100 \rangle$ to $\langle 111 \rangle$ with an increase in the overheating at the interphase crystal–melt boundary. X-ray diffraction analysis and microhardness determination were conducted to identify these intermetallic compounds. As a result of these studies, we established that the crystals have the chemical composition Al_6Mn and the measured H_μ value (3850 MPa) agrees with microhardness data obtained for the Al_6Mn phase by other authors [32].

The effect of HTT of melts on the structure of rapidly cooled alloys, in particular on the formation of aluminide crystals, was studied using Al–8% Fe and Al–2.5% Mn–2% Cr alloys. Below we set forth the most interesting results of these studies.

As follows from the data obtained, the phase constitution of the Al–8% Fe alloy remains constant over a wide range of cooling rates (from 10^2 to 10^4 K/s) and temperatures (from 1270 to 1620 K, at the melt overheating 150–450 K). In parallel with aluminides having the Al_3Fe composition and identical crystal lattices (an orthorhombic lattice belonging to the space group C_{mcm}), the Al–Al_3Fe eutectic forms in this alloy. However, variations in

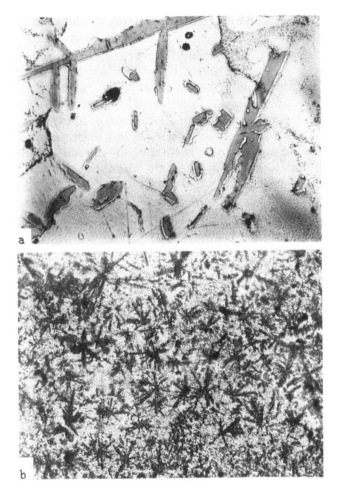

Figure 2.21   Growth forms of iron aluminide crystals in Al–8% Fe alloy at different solidification parameters: (a) $V = 1$ K/s, $\Delta T = 70$ K; (b) $V = 10^3$ K/s, $\Delta T = 470$ K ($\times 1200$).

the size and morphology of primary crystals are observed at $V > 100$ K/s. For example, the transition from the facet to dendritic growth forms occurs at $\Delta T = 200$ K (Fig. 2.21a) and a further rise in the melt temperature ($\Delta T > 350$ K) at $V = 10^3$ K/s causes the transformation of primary crystals into spherulite-like crystals (Fig. 2.21b, region III in Fig. 2.16d).

A general trend in the change of the size of iron aluminide crystals at various V with T is shown in Fig. 2.15d. Strong dispergation of the structure was found in castings obtained at $V > 10^4$ K/s. In this case spherulites are the preferable form of crystal growth in the whole investigated temperature range and their size at $\Delta T \leq 400$ K becomes comparable with the dimensions of eutectic crystals (the hatched area in Fig. 2.16d). We should emphasize that the given melt overheating lies in the temperature region close to T_{hom}.

A further increase in the cooling rate to $V = 10^5$ K/s at $T > 1470$ K and $\Delta T > 300$ K gives rise to a situation in which aluminides of different chemical composition (Al_6Fe) with orthorhombic structure ($a = 0.6492$ nm, $b = 0.7437$ nm, and $c = 0.8788$ nm) form in a rapidly cooled Al–8% Fe alloy in place of Al_3Fe crystals. The nature of this phenomenon will be considered below in greater detail.

These regularities in the transformation of growth forms of iron aluminides were also confirmed in the case of rapidly solidified alloys containing manganese aluminides. It was shown that the transformation of the growth forms typical of crystals of such compounds was observed with increasing V and T: faceted crystals, dendrites, and spherulites.

One of the most important peculiarities of the formation of primary intermetallic phases is the fact that faceted growth remains stable only in a limited range of solidification conditions, whereas a rise in melt temperature, as well as a growth in cooling rate, causes a transition to other mechanisms of crystal growth, which results in the observed diversity of branched crystals – dendrites and spherulites. Moreover, the use of HTT of melts may change the composition of aluminides. For example, aluminides in the specimens of an Al–2.5% Mn–2% Cr alloy obtained at low casting temperatures have a complex chemical composition and some portion of manganese atoms in their lattice is substituted for chromium. Overheating the melt promotes the transition of Cr to the α-solid solution and the primary aluminides formed in the melt have a composition and lattice corresponding exactly to the chemical compound Al_6Mn.

Thus, the effect of HTT of melts in forming aluminides with an orthorhombic lattice is revealed in the changes in their dimensions, shape, and growth mechanisms.

To conclude this section of Chapter 2, we formulate the main results of the experiment described in it and present some arguments to be used in the discussion based on the microheterogeneous model of the structure of aluminium melts:

• The stable form of crystal growth at low cooling rates and low melt overheating is the faceted form. The most extensive zone of faceted crystals forms in the growth of scandium aluminide, whereas the narrowest zone corresponds to the growth of iron aluminide.

• With a rise in cooling rate and melt overheating a transition is observed from faceted to rounded (dendritic, spherulitic, or globular) growth forms. The spherulitic forms are found in the process of solidification of aluminides Al_3Fe and Al_6Mn, while the globular forms are revealed in the formation of Al_3Sc aluminides and Si crystals.

• Melt overheating combined with high-speed solidification results in the formation of metastable phases Al_3Zr, Al_3Ti, and Al_6Fe.

It is known from general crystallization theory that the macroscopic shape of crystals is determined by the faceted interphase boundary and is related to the mechanism and kinetics of their growth, i.e. the morphological types of crystals of primary phases can be classified by the form of their growth as faceted or rounded [39, 40].

The aluminides considered above are high-entropy substances and rounded crystals is the main form of their growth. It is clear that aside from high entropy of melting, which determines the stable growth of rounded aluminide crystals in a wide range of solidification conditions, their faceting in growth is influenced by their crystallography. Because the above aluminides have different crystallographic lattices, it is easy to explain the observed diversity of their external faceting. For example, habitus in the form of cube or cuboctahedron is typical of the scandium aluminide and metastable zirconium and titanium aluminides, whereas crystals with lattices having a large crystallographic anisotropy grow largely as prolonged plates or needles.

It is known that the transition from a smooth to a rough interphase boundary occurs at the crystallization front with an increase of supercooling, and the tangential growth mechanism changes for the normal one, which results in the substitution of dendritic crystals for faceted ones [26, 29].

The appearance of dendritic growth forms upon solidification of alloys is usually dictated by the morphological instability at the crystallization front due to the thermal and concentration supercooling of the melt [26, 28]. Not going into details of dendritic growth in alloys, we can indicate a series of works published in the recent decades in which the theory of alloy solidification, explaining correctly the dendrite–grain structure formation in real alloys by the processes occurring in the two-phase zone, i.e. in the transition region between the liquid and solid phase, gained further development [29, 41].

The authors of the works [41–43] studied macroscopic dendritic-type growth structures by computer simulation methods and showed the evolution of dendritic structures in a wide range of supercooling, as well as the change of the dendritic form of growth of fcc crystals for the globular form, which is observed under extremely deep supercooling. Thus, the formation of such crystals in the process of growth of scandium and silicon aluminides and metastable zirconium and titanium aluminides revealed in our studies is not accidental, but is accounted for by the high symmetry of their crystal lattices.

The experimental investigations of solidification for strongly supercooled melts showed that with increasing driving force of the phase transition, dendritic crystals may be transformed into spherulitic crystals [44–46]. In these works a single mechanism of the formation of spherulitic crystals is verified both for metallic systems and for highly anisotropic substances such as selenium and tellurium. However, comparing the specific features of spherulitic growth in various systems, we can conclude that substances with more anisotropic lattices display a greater tendency to form such growth patterns. It is possible that this is one of the reasons for the formation of spherulitic crystals in the growth of iron and manganese aluminides.

When analyzing the joint influence of cooling rate and overheating of the melt on the morphological features of the growth of aluminides, the following experimental findings are important, in our opinion. From the schemes presented in Fig. 2.16 we see that at low overheating $\Delta T = 100$–150 K, usually employed in practice, the effect of the cooling rate reduces to the emergence of dendritic growth forms, which are observed at $V = 100$ K/s for iron and manganese aluminides and at $V > 10^3$ K/s for scandium aluminides. No other growth forms (except spherulitic for Al_3Fe crystals) that would characterize non-equilibrium solidification conditions are revealed. If we accept the viewpoint of Miroshnichenko [30] suggesting that for aluminium alloys, regardless of their composition and preparation conditions, the supercooling value δT at the crystallization front is determined by the cooling rate (for example, $\delta T = 25$ K at $V = 10^4$ K/s; $\delta T = 150$ K at $V = 10^5$ K/s), the observed sequence of shape formation for aluminides proves to be justified.

A more diversified picture is observed in the case when overheating the melt is used as a parameter specifying conditions at the crystallization front. Turning back again to the schemes presented in Fig. 2.16, we can see that for all the systems an increase in ΔT favours the appearance of more non-equilibrium forms of crystal growth – spherulites and globules. This effect may be caused only by additional melt supercooling.

To elucidate the nature of this phenomenon we traced the interrelation between the alloy structures in the liquid and solid states.

According to the experimental data described in Chapter 1 and in Section 2.1, liquid alloys of aluminium with transition metals are characterized by a clear-cut microheterogeneous structure. Near the liquidus temperature they consist of microscopic regions of two types differing by the content of the major components and the type of short-range order. One of the structural constituents is enriched with aluminium and retains the short-range order inherent in the alloy matrix; the second constituent is enriched with the transition element and its short-range order is similar to the structure of the original intermetallic phases. A rise in the melt temperature to a value specific for each alloy does not eliminate its microheterogeneity, whereas heating within this region promotes changes in the size, amount, and composition of the melt structural constituents.

Based on the classical approaches to the problems of crystal nucleation and growth [24–29], as well as on the recent works considering the nucleation processes at the level of association of clusters (or other microgroups with crystal-like atomic packing) [22, 23, 48, 50], we can assert that the formation of solidification centres in the melt is facilitated by a microheterogeneous structure [33–35].

A transformation of the melt structure related to an increase of the heating temperature should result in changes in the distribution of crystallization centers, their sizes and growth rate, and should control the mass transfer driving force at the crystal–melt interphase boundary. Assuming, in addition, that the binding energy of atoms in colloidal particles inheriting the structure of initial aluminides is proportional to the melting temperature of these particles and therefore decreases in the sequence Al_3Zr, Al_3Sc, Al_3Ti, Al_3Fe, Al_6Mn, the tendency of melts to supercooling will be determined by their structural constituents. Consequently, the melts alloyed with Mn and Fe are more prone to supercooling than the melts containing Sc or Zr. Therefore, the chance that such non-equilibrium forms of growth as spherulites will form in the solidification of these melts increases.

Of special interest, in our point of view, is the result (obtained for the first time) showing that the high-temperature heating of the melt ($T \leq T_{hom}$), when combined with high-speed solidification, gives rise to a substantial reduction in the size of structural constituents of the melt [33, 35].

The possible causes of this phenomenon are also related to the special condition of melts, which arises in overheating them to temperatures near T_{hom}. This condition is characterized by a decrease in the interphase tension at the boundaries of colloidal particles to the values at which the process of spontaneous dispergation of these particles begins (this effect was discussed in Section 2.1). This assumption is supported indirectly by the anomalous rise in viscosity on heating these melts, which is reflected in the character of its temperature dependences. A great number of nuclei of the solid phase arise in the process of subsequent solidification of this melt.

Of special interest and importance from a scientific and practical point of view are the results that give evidence of the formation of metastable aluminides during non-equilibrium solidification of Al–Zr, Al–Ti, and Al–Fe melts [33–35]. The formation of similar phases in Al–Zr and Al–Ti alloys was also observed in [36–38, 47] in the process of rapid solidification with cooling rates of 10^4 K/s and higher (the casting temperature did not exceed T_L by more than 100 K). The authors of these works suggested a mechanism of solidification based on the metastable phase diagram as a possible reason for the phase formation. However, the dependences of stability of the regions of growth for aluminides of this modification on the HTT of the melt obtained by us suggest that these regions are formed by another mechanism, namely, they are due to the change in the short-range order

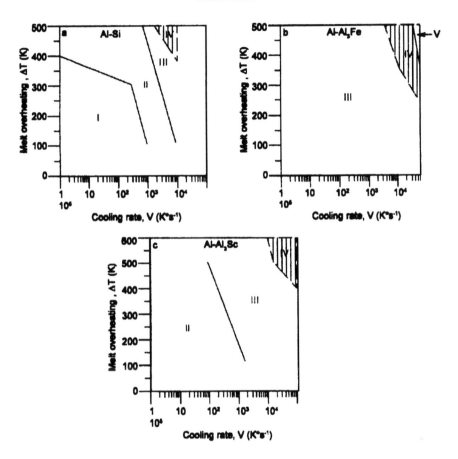

Figure 2.22   Various eutectic types in alloys: (a) Al–17% Si; (b) Al–8% Fe, (c) Al–2% Sc (I, plate; II, needlelike; III, globular; IV, globular divorced; V, metastable globular divorced).

in disperse particles of the microheterogeneous melt, which inherit the structure and composition of the compound.

Joint consideration of our experimental data and those reported in the literature [48–50] suggests that at a definite overheating of the melt, polymorphic-type transformations occur in disperse microgroups enriched with transition metals (this effect was discussed in Section 1.5). Since no evidence of such a transition in massive specimens at normal pressure was obtained, its reason may be either the size factor [51] or pressure [52] (the Laplace constituent of pressure caused by the interphase tension at a strongly curved boundary). As a result of such polymorphic transformation of the $D0_{23}–L1_2$ or $D0_{22}–L1_2$ type, microheterogeneities with a different short-range order form in the melt; they serve as potential solidification centres for metastable aluminides. High cooling rates are necessary in this case only for the fixation of such structural formations in the solid state. In conclusion it should be emphasized that the use of HTT of melts allowed us to extend the boundaries of the region of formation of the given phases toward lower cooling rates. We believe that the simultaneous presence of metastable and stable aluminides in the structure of rapidly cooled alloys (Fig. 2.18b) is further proof of the coexistence in the overheated Al–Zr melt of two types of colloidal particles differing by the structure of short-range order.

2.2.3. Some features of solidification of eutectics

Alloys with simple eutectics

Typical representatives of alloys with simple eutectics are the alloys of the Al–Si system, which constitute the basis for a group of silumins – casting aluminium alloys commonly used in practice. Taking into account the existing experimental material on the study of properties of liquid silumins, these alloys can be considered as an excellent model material for experiments aimed at the study of the role of HTT of melts in forming the structure of the solid phase [5, 31].

It is known that under equilibrium conditions the irregular eutectic – α-solid solution and silicon – forms at slow cooling of castings of silumins [53, 54]. Let us consider the morphological peculiarities of the eutectic formation in an Al–17% Si alloy as a function of solidification conditions in the range of cooling rates from 10 to 10^5 K/s, taking into account the overheating of the melt above T_L in the region of its microheterogeneous state.

The experiments showed that the melt solidification conditions determine the type of the eutectic and its properties by modifying the morphology and the size of the leading phase – crystals of the eutectic silicon.

Figure 2.22a shows schematically the succession of eutectic types with increasing V and ΔT. Not going into details of these changes, we can note that at $V < 10^3$ K/s silicon in the eutectic has a plate-type shape; at 10^3 K/s $< V < 10^4$ K/s the plates are changed for more disperse needle-shaped formations, whereas at $V > 10^4$ K/s a structure modification takes place due to the appearance of globular silicon crystals. The effect of melt overheating can be related to the further dispergation of silicon crystals observed for all types of eutectics.

For example, Figs. 2.23a,b show the influence of the melt temperature on the formation of plate-type eutectics. The effect of structure refinement is illustrated in the region of inception of the colony, where the regions of regular eutectic are retained. In the region of a globular eutectic the melt overheating $\Delta T > 350$ K causes a threefold refinement of silicon crystals and their size reaches 5 to 7 μm. The characteristic morphological feature of this structure is the presence of Si particles of nearly spherical shape in the interdendritic space of the α-phase (region IV in Fig. 2.22a).

Each of the above eutectic types may be characterized by an average microhardness value. The plate-type eutectic has $H_\mu = 550$ MPa; for the needlelike eutectic, $H_\mu = 850$ MPa; and for the globular eutectic $H_\mu = 1100$ MPa.

Eutectic systems with chemical compounds

Experimental data concerning the effect of HTT of melts on the structure and properties of eutectics with chemical compounds are lacking in the literature, although there is enhanced interest in such systems. This interest became the reason for conducting the experiments described below.

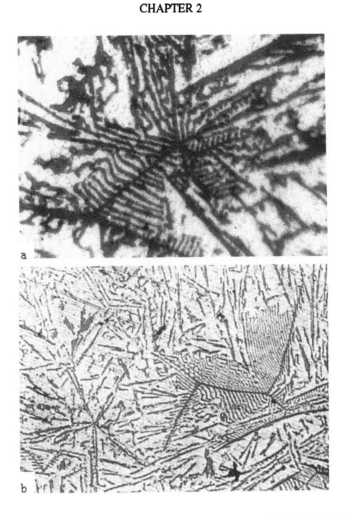

Figure 2.23   Influence of HTT of melt on the eutectic structure in Al–17% Si alloy ($V =$ 100 K/s): (a) $\Delta T = 50$ K; (b) $\Delta T = 300$ K ($\times 500$).

Regularities in the structure formation in eutectic systems with chemical compounds were considered by us in the alloys of hypereutectic compositions Al–8% Fe and Al–2% Sc [16, 33].

In rapidly cooled Al–Fe alloys the eutectic formed by the α-solid solution and Al_3Fe intermetallic compounds can be classified by its morphological features as anomalous and not continuous, in that eutectic Al_3Fe crystals have a nearly globular shape. Such a solidification pattern is retained in a wide range of cooling rates of the alloy and depends slightly on its composition.

The results presented in Fig. 2.22b show that heating the melt in the region above the liquidus to temperatures $T < T_{hom}$ significantly affects the structure formation in the solidification process, particularly at elevated cooling rates.

The most interesting features in the structure formation of Al–8% Fe, accompanied by a dramatic increase in H_μ up to 1100 MPa, were revealed under strongly non-equilibrium

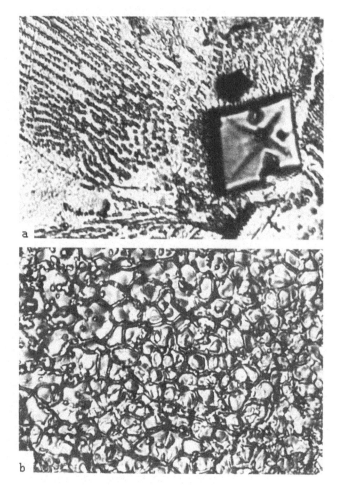

Figure 2.24 Various types of the Al–Al$_3$Sc eutectic in Al–2% Sc alloy as a function of solidification parameters: (a) $V = 1$ K/s, $\Delta T = 200$ K; (b) $V = 10_5$ K/s, $\Delta T = 500$ K ($\times 1200$).

solidification conditions, specifically, at $V = 10^4$ K/s and $T > 1573$ K ($\Delta T > 350$ K), and at $V = 10^5$ K/s in the temperature region 1473 K $< T <$ 1573 K ($200 < \Delta T < 300$ K).

It was established that the reason for improving material properties was the formation of a homogeneous eutectic structure with Al$_3$Fe aluminides of size 1 to 2 μm (region IV in Fig. 2.22b).

A relatively high degree of overheating of the melt ($\Delta T > 250$ K), in combination with the cooling rate $V = 10^5$ K/s, results in the formation of another type of eutectic in Al–Fe alloys (region V in Fig. 2.22b). In this case, intermetallic compounds of the metastable phase Al$_6$Fe are formed in place of Al$_3$Fe aluminides; they also have an orthorhombic lattice but with the lattice parameters $a = 0.6492$ nm and $b = 0.7437$ nm differing from those in the Al$_3$Fe crystal structure.

A similar structure was also obtained for another eutectic with intermetallic compound, namely, for Al–Al$_3$Sc (region IV in Fig. 2.22c). The conditions for its formation in the

Al–2% Sc are: $V = 10^4$ K/s and 1570 K $< T <$ 1670 K. It is evident that in this case a continuous needlelike eutectic, which is the typical growth form at $V < 100$ K/s (Fig. 2.24a), degenerates and a eutectic consisting of disperse Al_3Sc crystals surrounded by equiaxed grains of the α-solid solution is formed (Fig. 2.24b). This indicates that under such solidification conditions significant changes are observed not only in the morphology of the eutectic structure but also in the mechanism of its growth.

It is difficult to explain the formation of such anomalous structures by simply invoking the "classical" eutectic growth theories [54]; however, the picture will become clearer if the concept of the microheterogeneous state of metallic melts is used [16, 31, 33].

All the alloys considered here have a hypereutectic composition and the eutectic formation occurs in these alloys after the nucleation of crystals of primary phases, thereby affecting the specific features of eutectic growth. The most typical manifestation of this effect is readily proved by the structures obtained in the slow solidification of the Al–17% Si alloy (Figs. 2.23a,b). An important role in the formation of such complex regular structures belongs to the concentration supercooling produced ahead of the crystallization front. When a hypereutectic alloy solidifies, there arises a one-phase instability of the eutectic phase boundary, resulting in the leading growth of the faceted phase and in the formation of eutectic cells or dendrites after the formation of the diffusion boundary layer. The shape of these cells is determined by the value of concentration supercooling, crystallographic characteristics, heat removal, and surface energy. However, for a given material, the first factor plays by far the largest role.

Our results [16, 31, 33] and the data on non-equilibrium solidification of eutectics [30, 53–55] show that the dispergation and the change in morphology of the eutectic constituents occur with an increase in the cooling rate. These changes are related to the increase in supercooling with the eutectic growth. According to Taran and Mazur [53], rapid solidification of an Al–Si eutectic is accompanied, by virtue of the kinetic factors, by a change in the relation between the rates of growth of the constituent phases: the phase with lower entropy of fusion – the aluminium phase – becomes the leading one.

Using the data of metallographic examination, we can say that such morphology of eutectics was observed by us for all compositions considered – for the Al–Si eutectic in region III (Fig. 2.22a), for the Al–Al_3Sc eutectic in region III (Fig. 2.22c), and for the Al–Al_3Fe eutectic in region IV (Fig. 2.22b). In all cases the eutectic solidification was preceded by the primary crystallization of the corresponding aluminides (cf. Figs. 2.16 and 2.22), which imposed limitations on the conditions of nucleation of the secondary phase of the same composition.

A special relation between these phases is established, in our opinion, under definite solidification conditions corresponding to the hatched areas in Fig. 2.22. A simple comparison of the parameters involved shows that in all cases such a situation arises when a high cooling rate is combined with strong melt overheating. As was indicated earlier, because of the specific structure of liquid alloys of aluminium with transition metals their heating to $T \leq T_{hom}$ and subsequent rapid quenching do not suppress the growth of primary aluminides but, on the contrary, increase their number, abruptly reducing the size of crystals (to 1–2 μm). The initially growing crystals produce specific conditions for the formation of analogous eutectic phases which arise and develop on these crystals as on supports. The second eutectic phase, the α-solid solution, occupies the remaining volume and grows independently of the first phase. According to the commonly accepted terminology, such a eutectic is called a degenerate divorced eutectic, which means that a single phase boundary is lacking during its growth. The relationship between the phases in the eutectic remains unchanged in this case and the total amount of aluminides (or silicon) in such a structure is added from primary and eutectic phases, which cannot be distinguished by morphology and sizes.

Thus, the appearance of divorced eutectics is accounted for, in our opinion, by the specific structure of aluminium melts alloyed with transition metals or silicon, and proves once again the interrelation between the structures in the liquid and solid states.

2.3 Peculiarities of the Structure Formation in Alloys Obtained with the Use of Homogenizing HTT of Melts

As pointed out in Chapter 1 and Section 2.1, the conversion of a melt into the homogeneous state considerably changes its structure and properties. Much evidence in favour of the common behaviour of liquid and solid alloys suggests that the use of the homogenizing HTT of melts can radically transform the structure of the material.

In this section we describe the results of the study and properties of materials obtained with the use of HTT of melts in the temperature region above T_{hom}. To reveal the regularities of solidification after such a treatment we considered alloys with different phase diagrams [5, 16, 17, 31].

Specimens of the compositions indicated above were obtained in the form of ingots ($V = 1$ K/s), permanent moulds ($V = 100$ K/s), and ribbons ($V = 10^5$ K/s). The homogenizing HTT involved heating to temperatures exceeding T_{hom} by 100 K.

Table 2.3   Compositions of alloys under study.

Phase diagram type	Composition, %	Ref.
Simple eutectic	Al–17% Si Al–26% Si	[31]
Eutectic with intermetallic compound	Al–2% Sc	[16]
Peritectic	Al–0.6% Cr Al–5% Cr	
	Al–4.5% Ti	[34]
	Al–0.6% Zr	
	Al–2% Zr	[10]
Monotectic	Al–In, Al–In	[17]

2.3.1. Alloys with eutectic phase diagram

Let us consider the results obtained on the simple Al–Si eutectic.

Figure 2.25 shows microphotographs of such structures. Whilst having a high silicon content, the Al–17% Si alloy does not contain precipitates of primary phases, i.e. the

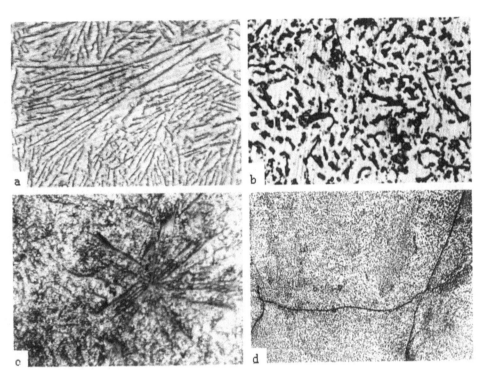

Figure 2.25   Various types of quasieutectic structures as a function of composition and parameters of solidification of alloys: (a) Al–17% Si, $V = 10$ K/s, $T > T_{hom}$; (b) Al–17% Si, $V = 10^3$ K/s, $T > T_{hom}$ ($\times 200$); (c) Al–26% Si, $V = 10^5$ K/s, $T > T_{hom}$ ($\times 2000$); (d) Al–2% Sc, $V = 100$ K/s, $T > T_{hom}$ ($\times 2000$).

structure consisting of long thin platelets of the eutectic Si and the aluminium-based α-solid solution is a quasieutectic formation (Fig. 2.25a). This eutectic is typical of specimens obtained at low cooling rates and corresponds to the plate type. An increase in the cooling rate changes the crystal morphology of eutectic silicon and, correspondingly, the plate type of eutectic for needlelike (Fig. 2.25b). It is evident that suppression of the primary crystallization of silicon changes the proportion of phases in the eutectic in favour of the silicon eutectic phase. This is evidenced by data from the qualitative metallographic examination of the structure of permanent mould samples and the fact that their microhardness increased by 300–350 MPa.

The structure of ribbons obtained from the homogeneous melt undergoes even greater changes: primary dendrites of the α-phase were detected against the background of globular eutectics, i.e. the alloy crystallizes under such conditions as a hypoeutectic one.

Thus, homogenizing HTT of melts may be thought of as promoting deeper supercooling at the crystallization front and because of this the stable equilibrium phase diagram of the alloy transforms into a metastable diagram at even ordinary solidification rates. The combination of high cooling rates with the homogenizing HTT leads to a further change in the phase diagram, which manifests itself in the appearance of primary crystals of the α-solid solution in alloys with initially hypereutectic composition.

Metallographic measurements of volume relations of the structural constituents in ribbons of an Al–26% Si alloy showed that after homogenizing HTT of the melt (at $\Delta T = 500$ K) the fraction of crystals of primary silicon is only 4–5%, i.e. a factor of 4 lower than that prescribed by the equilibrium phase diagram. An increase in ΔT to 700 K suppresses completely the growth of primary precipitates and leads to the formation of a quasieutectic structure consisting of branched silicon crystals similar in shape to spherulites and distributed uniformly in the aluminium matrix (Fig. 2.25c). The H_μ values for such a eutectic are twice as great (1100–1200 MPa) as those for the equilibrium eutectic.

Thus, an increase in the Si content in the alloy leads to the rise of the absolute value of the melt overheating temperature at which the stable growth of a quasi-eutectic is observed. Moreover, the formation of such a specific form of growth of the solid phase as a spherulitic eutectic colony is related to the instability of the plane crystallization front arising at strong concentration and thermal supercooling, which is caused in turn by the specific condition of the melt after its homogenization.

The universal character of this phenomenon is corroborated by the results obtained for more complex eutectic systems, for example, for a eutectic system with the chemical Al_3Sc compound [16]. According to the equilibrium phase diagram, the scandium concentration at the eutectic point is 0.6%. In our experiments we examined the structure of ingots prepared from an Al–2% Sc alloy with a hypereutectic composition at equilibrium conditions. As follows from Fig. 2.25d, overheating the melt to a temperature exceeding the temperature region of the melt microheterogeneity, i.e. above 1660 K, made it possible to suppress the primary crystallization of intermetallic compounds and obtain ingots with the modified globular quasi-eutectic consisting of eutectic Al_3Sc crystals uniformly distributed in the aluminium phase.

Thus, the established regularities in the structure formation of eutectic aluminium alloys crystallized after overheating the melt to $T > T_{hom}$ show that the tendency of the melt to supercooling abruptly increases upon its homogenization.

Because of the specific structure of the given eutectics an increase in melt supercooling will promote phase nucleation in the zone of joint eutectic growth (i.e. in the temperature–concentration part of the phase diagram where the combined growth of eutectic phases occurs at higher rate than their isolated growth) [54]. For eutectics with phases having widely different entropies of fusion, this zone is displaced toward higher concentration of the phase with higher entropy, i.e. to silicon and scandium. Occurrence in this zone provides a higher rate of growth of the low-entropy phase, making it the leading phase.

From the above reasoning it can be concluded that eutectics containing a single high-entropy phase may solidify at sufficiently high supercoolings (caused by the increase in the cooling rate or by overheating the melt to the temperatures $T > T_{hom}$) with an increased volume fraction of this phase compared to the equilibrium phase. This fact may explain the increase in microhardness by 300–350 MPa in the Al–17% Si alloy and by 550–600 MPa in the Al–26% Si alloy.

In our opinion, the most important result is the formation of a quasieutectic structure in Al–Si and Al–Sc alloys at low cooling rates ($V < 10$ K/s) after high-temperature homogenizing treatment of the melts.

It was thought in [30, 53] that suppression of the primary solidification and the growth of quasieutectics are possible only at high cooling rates, because such rates create the

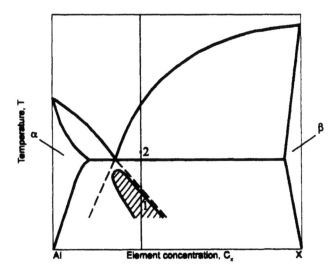

Figure 2.26   Schematic view of the zone of joint growth of eutectic phases and the regions of formation of quasieutectic structures (1) and divorced eutectics (2) in alloys of hypereutectic composition.

necessary non-equilibrium conditions at the crystallization front. Analysis of this experimental material shows that such a viewpoint existed because the authors ignored the existing relation between the liquid and solid states of the alloys and did not overheat the melts above T_L by more than 200 K.

We can suggest a more correct treatment of these results. In our opinion, at high V and insignificant ΔT the divorced eutectic forms in alloys with anomalous eutectics instead of the quasieutectic. It is easy to make sure that the distinction between these two structures lies in the kinetics of the phase formation, whereas the morphology and microstructure are entirely identical.

To make the proposed model more obvious we present in Fig. 2.26 a schematic view of part of the phase diagram with the zone of joint eutectic growth where the conditions for the formation of quasieutectics (1) and divorced eutectics (2) are shown. In spite of the conditional character of this scheme, distinctions in the mechanisms of formation of such structures are evident.

When considering the possible variants of formation of the quasieutectic structures in alloys related to the high-temperature treatment, we should take into account the width of the solidification range. It is quite clear that the smaller its value, the lower the cooling rates required to get into the zone of joint growth and to form the quasieutectic structure. For example, in Al–Si alloys the range of solidification increases from 90 to 200 K as the silicon concentration increases from 17 to 26%. As a consequence, in spite of the homogenizing treatment in the liquid state, the quasieutectic structure is lacking at low cooling rates in the Al–26% Si alloy.

Thus, the results described above show once again that supercooling the melt is a major external factor acting upon the genetic features determining the formation and morphology of eutectic structures.

2.3.2. Alloys with peritectic phase diagram

The results described in Section 2.2 showed that peritectic systems are highly sensitive to the effect of HTT of melts. This manifests itself in the possibility of controlling, throughout a wide range, the morphology and kinetics of growth of the phase constituents of alloys even when the temperature of heating the liquid metal did not exceed T_{hom}.

In this section we present experimental data on the formation of the solid phase in the process of solidification of liquid binary and ternary aluminium alloys with Cr, Zr, and Ti, which were preliminarily overheated to $T > T_{hom}$. Chill castings and ribbons of 300–400 μm thick are examined. The structures of samples melted at different HTT of melts (heating to $T < T_{hom}$, $T = T_{hom}$, and $T > T_{hom}$) are compared.

The preparation conditions and the results of X-ray diffraction analysis of the samples made of Al–Cr alloys are presented in Table 2.4.

Table 2.4 Composition and solidification conditions for alloys of the Al–Cr system.

Cr content, %	Melt heating temperature T, K	Cast temperature T_c, K	HTT	Phase composition
0.6	1340	1340	$T < T_{hom}$	α-solid solution + Al_7Cr
	1490	1490	$T > T_{hom}$	Oversaturated α-solid solution
0.6	1490	1210	$T > T_{hom}$	Oversaturated α-solid solution
0.6	1490	1120	$T > T_{hom}$	Oversaturated α-solid solution
5	1320	1320	$T < T_{hom}$	α-solid solution + Al_7Cr
5	1520	1520	$T > T_{hom}$	Oversaturated α-solid solution
5	1520	1220	$T > T_{hom}$	Oversaturated α-solid solution

The structural feature shared by the samples of both compositions melted at $T > T_{hom}$ is the suppression of the growth of primary intermetallics and formation of the aluminium-based supersaturated α-solid solution. It is important that this one-phase condition is retained even in the case when the alloy overheated above T_{hom} was cooled down to $T_L < T < T_{hom}$. After such a treatment, the character of solidified grains of the anomalously supersaturated α-solid solution remains dendritic and the size of the dendritic cell reduces by a factor of 1.5–2.

The results of metallographic examination, X-ray diffraction and X-ray spectral studies of the structure of ribbons prepared from the Al–0.6% Zr alloy are presented in Table 2.5 and in Figs. 2.27 and 2.28.

As follows from Fig. 2.27, ribbons cast from the melt overheated to $T = 1420$ K (dark dots) have a more disperse structure than ribbons cast from the melt overheated to the higher temperature ($T = 1490$ K, bright dots). Thus, in spite of the same casting

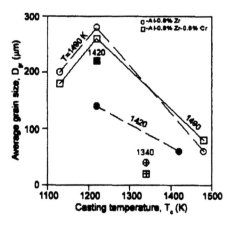

Figure 2.27 Average grain size versus temperature of casting of ribbons obtained from Al–0.6% Zr and Al–0.6% Zr–0.6% Cr alloys after different HTT regimes in the liquid state.

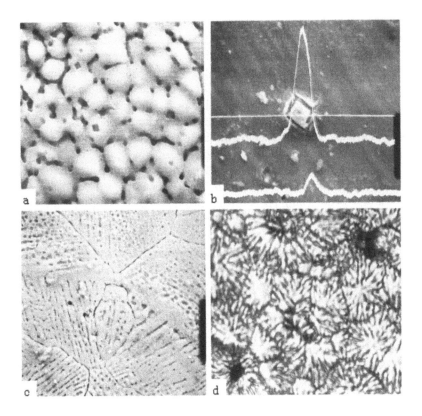

Figure 2.28 Structure of ribbons of Al–0.6% Zr (a, b, c) and Al–4% Ti (d) alloys obtained after different regimes of HTT of their melts: (a, b, c) backscattered electron image and zirconium X-ray K_α image; (d) optical micrograph, $\times 2000$. a, b, $T < T_{hom}$; c, d, $T > T_{hom}$.

Table 2.5   Conditions for producing ribbons from Al–0.6% Zr alloy.

T, K	T_c, K	ΔT, K	HTT
1340	1340	370	$T \ll T_{hom}$
1420	1420	450	$T < T_{hom}$
1420	1220	450	$T < T_{hom}$
1420	1120	450	$T < T_{hom}$
1490	1490	520	$T = T_{hom}$
1490	1220	520	$T = T_{hom}$
1490	1120	520	$T = T_{hom}$
1520	1520	550	$T > T_{hom}$
1520	1220		$T > T_{hom}$

temperature, the structures of ribbons prepared from the melts subjected to different HTTs are quite different.

Metallographic examination of the structures of ribbons that were cast at various regimes points to a decrease in the amount of primary intermetallics of the metastable Al_3Zr phase located at the centres of modified grains of the α-solid solution and to the coarsening of these grains with increasing cast temperature (Figs. 2.28a,b). Overheating the melt into the region of temperatures close to T_{hom} changes the phase constitution of the melt, and as a result a metastable, one-phase, anomalously supersaturated α-solid solution (with $a = 0.40507$ nm) forms in place of the equilibrium heterogeneous two-phase structure (Fig. 2.28c). Additional holding of the melt at such a temperature, as well as cooling at lower cast temperature, does not lead to qualitative changes in the alloy structure.

Adding a third component – chromium – to an Al–0.6% Zr alloy in amounts of 0.6% did not cause significant alterations in its structure formation and merely increased the degree of alloying the α-solid solution, which is evidenced by the reduced lattice parameter of the α phase ($a = 0.40475$ nm). The formation of anomalous structures in the ternary alloy was observed at the same HTT regimes.

The positive role of the homogenizing HTT of a melt in creating one-phase structures of anomalously supersaturated α-solid solutions with transition metals is confirmed by the results obtained when solidifying an alloy with higher Zr content. For example, for the Al–2% Zr alloy such a structure is formed in ribbons obtained by rapid quenching at $V = 10^4$ K/s. Meanwhile, if the melt was not subject to the preliminary HTT, the suppression of primary solidification of aluminides in the alloy of the given composition occurs in cooling with rates of the order of 10^5 K/s.

Thus, the use of homogenizing HTT of melts in solidification of alloys of aluminium with transition metals extends the region of existence of anomalously supersaturated Al-based α-solid solutions due to the suppression of the growth of primary aluminides and the formation of the one-phase state at lower cooling rates.

The structural regularities described above were also detected on solidification of Al–Ti alloys of hyperperitectic composition.

To ensure melt homogeneity conditions, a liquid alloy (Al–4.5% Ti) was preliminarily overheated above T_L by more than 400 K. When the melt solidified at $V = 10^5$ K/s, the one-phase structure of the α-solid solution with lattice parameter $a = 0.40415$ nm formed in the ribbon structure, which attests to the anomalous supersaturation of the solid solution. Another factor which proves that solidification occurred under non-equilibrium conditions is such morphological characteristic as the spherulite-like growth of grains of the α-phase (Fig. 2.28d).

The above results confirm the fact that solidification of a strongly overheated melt is implemented at significant deviations of the system from the equilibrium state.

All peculiarities of the ribbon solidification listed above were confirmed by solidifying chill casts. This suggests that the homogenizing HTT of melts is an effective means for changing the structure of the material obtained both at normal and increased cooling rates.

2.3.3. Alloys with monotectic phase diagram

Alloys of the Al–In system are typical representatives of the alloy class with a monotectic phase diagram. The existence in the subliquidus part of the phase diagram of a wide region of immiscibility of two liquids bounded by an immiscibility dome hampers the preparation of uniform materials on account of the enrichment of the bottom part of the ingot in the heavier component.

Ivanov *et al.* [56] could suppress the immiscibility of liquid by performing solidification under zero gravity conditions or in crossed electric and magnetic fields. A fairly uniform height distribution of components with precipitates of ~1000 μm in size was observed in this case. A macroscopically uniform structure with very fine precipitates in hypermono-tectic alloys can be formed as a result of solidification of liquid metal at a cooling rate of 10^3–10^5 K/s [2]. However, such solidification conditions can be realized in industrial technological processes only to produce metallic powders or thin ribbons, but not in large-scale production of massive castings.

Based on the positive experience obtained on eutectic and peritectic alloys, Popel *et al.* [17] studied the possibility of suppressing or slowing down macrolamination in massive samples of Al–In alloys with the aid of HTT of their melts. The temperature regime was chosen on the basis of the results of measurements described in Section 2.1. According to these data, the authors determined the critical temperatures, overheating above which destroys the microheterogeneous state of the melt existing upon heating beyond the limits of the immiscibility dome.

The effect of homogenization of melts on the structure formed in their solidification with moderate cooling rates ($V \sim 1$ K/s) was studied on 30 g samples melted in an open resistance furnace and containing 17, 20, 30, and 40% In. The melt was heated to the desired temperature, soaked for 30 min and cooled at the above rate. To reproduce a more accurate pattern of the structure of hypermonotectic melts, ribbons obtained by high-speed solidi-fication with a cooling rate of 10^5 K/s were studied.

The results of metallographic examination are summarized in Table 2.6.

It was found that for all compositions studied at melt heating temperatures below T_{hom} a boundary between the phases enriched with various components forms upon solidi-fication (Fig. 2.29a). X-ray phase analysis confirmed the separate existence of aluminium (at the top) and indium (at the bottom) phases. As the melt temperature rises and approaches T_{hom} in solidification, this boundary is smeared and further, at $T > T_{hom}$, complete suppression of macrosegregation in ingots occurs. In this case the structure is analogous to

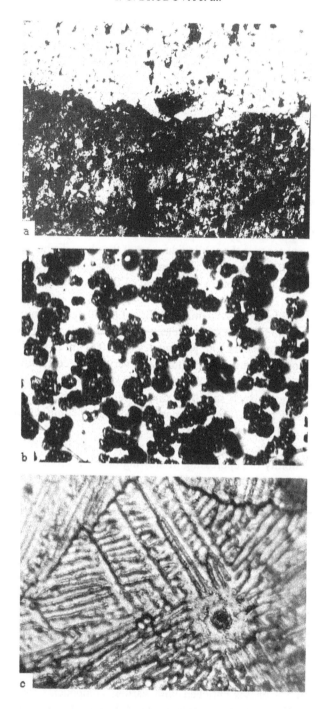

Figure 2.29   Structure of Al–20% In alloy at different solidification parameters: (a) $V = 1$ K/s, $\Delta T = 70$ K ($\times 80$); (b) $V = 1$ K/s, $\Delta T = 600$ K ($\times 500$); (c) $V = 10^5$ K/s, $\Delta T = 600$ K ($\times 1200$).

Table 2.6   Conditions for specimen preparation from Al–In alloys.

Indium content, %	T, K	HTT of melt	Presence of macrosegregation
17	970	$T < T_{hom}$	yes
	1270	$T < T_{hom}$	yes
	1450	$T = T_{hom}$	no
20	970	$T < T_{hom}$	yes
	1270	$T < T_{hom}$	yes
	1470	$T = T_{hom}$	no
	1570	$T > T_{hom}$	no
30	1270	$T < T_{hom}$	yes
	1470	$T < T_{hom}$	yes
	1520	$T < T_{hom}$	yes
	1570	$T = T_{hom}$	no
	1620	$T > T_{hom}$	no
40	1470	$T < T_{hom}$	yes
	1570	$T < T_{hom}$	yes
	1670	$T = T_{hom}$	no
	1720	$T > T_{hom}$	no

the quasieutectic one (Fig. 2.29b) and consists of particles of the indium phase uniformly distributed throughout the volume of the aluminium matrix. A further rise in the melting temperature significantly influences the ingot structure, facilitating the refinement of precipitates of the disperse indium phase in a macroscopically uniform ingot.

Histograms of mean sizes of particles at various melt temperatures are presented in Fig. 2.30. It is evident that the number and diameter of indium particles sharply decreases on passing through T_{hom}, which makes the structure even more modified. In this case the lines relating to the two phases – aluminium solid solution and indium – are present in the diffraction pattern. To characterize the aluminium-based solid solution, the parameter values of its crystal lattice were analyzed. This parameter was determined with allowance

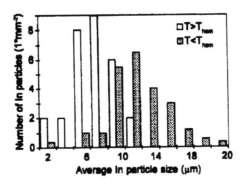

Figure 2.30   Distribution of the average size and the number of particles of the indium phase in the structure of Al–20% In alloy obtained under different regimes of HTT of melt.

for temperature correction as an average over five measurements using $(331)_\alpha$ and $(420)_\alpha$ reflections in the CoK_α radiation. No indications of supersaturation of the α-solid solution were found.

The ratio of components varies with alloy composition: the increase in the indium content to 30–40% raises the fraction of the indium phase but influences only slightly the dispersity of precipitates.

X-ray and metallographic studies of an Al–20% In alloy rapidly cooled at a rate of 10^5 K/s showed that even low overheating of the melt prevents macrosegregation and as a result the structure corresponding to a hypomonotectic composition forms. Grains of the solid solution of indium in aluminium are distinctly seen in this structure, with interlayers of the indium phase located at the grain boundaries. As the temperature of the melt rises, the dendritic character of the growth of crystals of the α-phase is enhanced, while the other, In-rich phase occupies the interdendritic space. On solidification of the melt preliminarily overheated to temperatures above T_{hom}, a structure exhibiting all the signs of spherulitic growth of crystals of the solid phase is revealed. This type of structure is shown in Fig. 2.29c. The structure elements grow from a single center (a single-crystal nucleus) and have a radial divergence in the heat flow direction. As was shown earlier, such nucleation and subsequent splitting of the spherulitic nucleus is observed at significant overheating conditions of the melt on account of the morphological instability of the crystal–melt interphase boundary. As the solid phase grows and overheating decreases, the spherulite-like grains develop by a dendritic mechanism keeping the characteristic rounded shape due to radial heat removal.

The results obtained show that homogenizing overheating of the melt slows down its macrosegregation in subsequent cooling. Rapid segregation of samples that were not over-heated is due to the preservation in liquid metal of In-rich colloidal particles, which play the role of nuclei of the indium phase upon cooling under the immiscibility dome. These nuclei are lacking in the homogenized melt and consequently its decomposition begins at deeper supercooling and proceeds less intensively. If the quenching rate is rather high, the melt retains a high degree of homogeneity up to the liquidus line. This promotes melt supercooling before solidification and the formation of primary crystals of the α-phase with a spherulitic shape. The change in the dispersity of indium precipitates with increasing the temperature of melt overheating above T_{hom} indicates that the colloidal structure is not completely destroyed on passage through T_{hom} and lower-scale microheterogeneities are still retained in the melt, whose nucleation effect manifests itself only at deeper super-cooling.

Thus, in alloys of monotectic systems solidified from a melt subjected to preliminary homogenizing treatment the tendency to macrosegregation declines, which allows one to obtain castings with more uniform structure and properties.

2.4 Commercial Alloys Produced with the Use of Technology Based on HTT of Melts

In this section the wide opportunities of the effective use of HTT of melts for governing the structure and properties in the solid state are illustrated by a series of granulated and casting alloys.

2.4.1. Granulated alloys

From a practical point of view, the most interesting materials obtainable by this method are heat-resistant alloys of the Al–Zr–Cr system.

According to previously known results [2], high heat resistance of these alloys is provided by high-melting additives which precipitate in the form of disperse intermetallic compounds in reprocessing granules or ribbons into semi-finished products. On this basis it can be concluded that the technology of producing semi-finished products from granulated alloys calls for a uniform fine structure in the as-cast condition. This task is complicated, however, by the presence of crystals of primary aluminides in the cast structure of alloys of the given compositions. The presence of such crystals, drastically impairing mechanical properties of the alloys, is accounted for by the low solubility of transition metals in aluminium. The most traditional method to eliminate them is the use of technologies of ribbon or powder production, in which the cooling rate in quenching from the liquid state reaches values of more than 10^4 K/s.

In our experiment for these purposes we used HTT of the melt as an alternative to the increase in cooling rate. The basic stages of this new technology were developed on specimens obtained with a high-speed solidification setup modelling conditions for casting granules of an Al–Cr–Zr alloy.

Ternary alloys with the total content of transition metals below 3% and above 3–4.5% were investigated separately. The HTT regimes were chosen on the basis of data obtained in the study of structure-sensitive properties of the melts. It was found that the behavior of melts of the Al–Cr–Zr system in the process of heating and cooling is similar to their behavior for the binary systems described in detail earlier. The T_{hom} temperatures determined for ternary alloys of the indicated compositions lie in the temperature range from 1500 to 1520 K.

Analysis of morphological features of the structural constituents of ternary alloys and X-ray diffraction study confirmed the results [2] showing that in this ternary system, irrespective to the chromium and zirconium content, only the binary Al_7Cr and Al_3Zr compounds are formed in addition to the aluminium solid solution.

As follows from the study of the structure formation in rapidly cooled binary Al–Zr alloys, Al_3Zr intermetallic compounds of metastable modification form with a zirconium content from 1 to 3% in the range of cooling rates 10^2–10^4 K/s.

In ternary alloys the formation of these aluminides was established under the same conditions; it was shown that the use of HTT of the melt enables one to alter their morphology and sizes over a wide range. The following results are of greatest interest. As melt overheating increases, the rounded cubic forms of growth change for dendritic forms, differing in the degree of anisotropy of the growth rate of secondary arms. In addition, preliminary heating of the melt to temperatures near T_{hom} increases sharply the amount of aluminides and reduces their sizes.

Primary crystallization of intermetallic compounds affects the formation of the major structural constituent of the melt – the α-solid solution. For example, according to the data reported by Ohashi and Ichikawa [47] and to the metallographic studies described here, grains form at precipitates of a metastable Al_3Zr phase. The argument in favour of the concepts stated here is the formation of a disperse, 5 μm-sized subdendritic structure in the specimens prepared from the Al–3% Zr–1% Cr alloy subjected to homogenizing HTT in the liquid state. Although the melt temperature was in excess of T_{hom}, the stable growth of fine metastable Al_3Zr aluminides was observed in the cast structure of the alloy. However, with a rise in the heating temperature of the melt additional alloying of the α-solid solution with zirconium took place, which is evidenced by the growth of the lattice parameter and microhardness (by 20 MPa).

For a more accurate estimate of the degree of supersaturation of the α-solid solution direct measurements of zirconium concentration were carried out with an electron probe microanalyzer. They confirmed that the use of HTT raises the zirconium content in the α-solid solution by 0.25–0.3%.

Thus, the use of HTT of melts in technologies with cooling rates lower than 10^4 K/s does not eliminate the formation of primary aluminides in the structure of alloys with a total content of chromium and zirconium of more than 4%.

A different pattern is observed on solidifying alloys with the total content of the components equal to 3%. In this case the range of stable growth of metastable Al_3Zr intermetallic compounds shifts toward lower cooling rates (10^3 K/s), while additional homogenizing HTT of the melt causes the formation of a dendritic structure of the anomalously supersaturated α-solid solution.

Summarizing the experimental results described above, we can conclude that two different HTT regimes can be used for improving the structure of granulated heat-resistant alloys of the Al–Cr–Zr system.

The first regime, with $T > T_{hom}$, enables one to suppress the primary crystallization of aluminides and obtain the structure of the anomalously supersaturated α-solid solution at cooling rates exceeding those required without HTT by an order of magnitude. Such a treatment can be effectively applied to alloys when the total content of the components is 3% and lower. An increase in the degree of alloying of the α-solid solution enhances its thermal stability and raises the strength properties. According to [1], such an alloy possesses the following mechanical characteristics: $\sigma_u = 380$ MPa, $\sigma_{0.2} = 340$ MPa, $\delta = 12\%$.

The second regime, with $T \leq T_{hom}$, provides the formation in the alloy structure of disperse crystals of the metastable phase Al_3Zr and the supersaturated α-solid solution. The presence of such aluminides provides an additional strengthening effect, with the maximum strength of the alloy achieved when the zirconium content exceeds its solubility limit (at the given cooling rate).

As an additional example of the positive influence of the HTT of melts on the structure and properties of rapidly cooled alloys we can adduce the experimental results obtained for alloys of the Al–Zn–Mg–Cu system with additions of transition metals – zirconium and chromium.

In spite of the explicit advantage of these alloys over traditional deformable alloys (such as a 7055-type alloy), semi-finished products made of these alloys have unstable mechanical properties and unstable technical strength characteristics. As was established, the cause of such instability is the primary zirconium aluminides of metastable modification arising in the structure of granules, which prevent, owing to their dendritic structure and the size of about 10 μm, the formation of a one-phase, anomalously supersaturated α-solid solution. To improve the structure of granules, an alternative regime of casting was suggested in which the temperature of preparation of the melt with high-melting additives was raised up to the temperature of the melt homogenization.

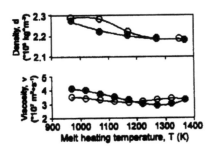

Figure 2.31 Temperature dependences of kinematic viscosity ν and density *d* of AL9 alloy in the liquid state (• heating, ○ cooling).

The results of metallographic and X-ray spectral studies showed that in this case we obtain the structure of the α-solid solution anomalously supersaturated with low-melting and high-melting elements, which provides stable and higher mechanical properties after heat treatment in the solid state. For example, the ultimate strength increased by 10%; relative elongation by 60%; and low-cycle fatigue and fracture toughness by 32 and 20%, respectively.

2.4.2. Casting alloys

Silumins, i.e. alloys of the Al–Si system, find wide application in the production of castings with high operating characteristics, such as tightness, corrosion resistance, and strength. A typical representative of this class of alloys is the AL9 (A356) alloy having a composition close to eutectic and sufficiently high strength characteristics due to magnesium alloying. However, the use of this alloy for producing castings of an intricate geometric shape calls for increased plasticity, while keeping the same level of strength characteristics.

To improve the quality of castings made of silumins, Brodova and Bashlikov [57] suggested the use of HTT of the melt as in the experiment described below. The HTT regimes were chosen on the basis of the results of investigations of structure-sensitive properties of melts, in particular, temperature dependences of viscosity and density. The results of these measurements are presented in Fig. 2.31. They confirm the fact that silumin melts are complex microheterogeneous systems retaining the metastable state up to comparatively high temperatures (1220 K [5]).

Two regimes of HTT of the melt were studied: with $T < T_{hom}$ and $T > T_{hom}$. The special technology of making castings from the AL9 alloy (6–8 Si, 0.2–0.4 Mn, 0.3–1.0 Fe, 0.2–0.4 Mg) in industrial conditions involved the following operations. Metal melted in a gas furnace was heated in an electrical furnace to the preselected temperature T, soaked for 30 min, and cast into a permanent metal mould, i.e. the cast temperature T_c was equal to T. In another batch of castings solidification conditions were different: after overheating the melt its temperature was lowered to the cast temperature ($T_c = 990$ K). Consider first the features of the formation of castings after preliminary HTT of the melt in the mode with $T < T_{hom}$. Two series of specimens were melted. In the first $T = T_c$ and T_c varied from 990 to 1130 K; for the second series T varied within the same limits, while T_c was 990 K.

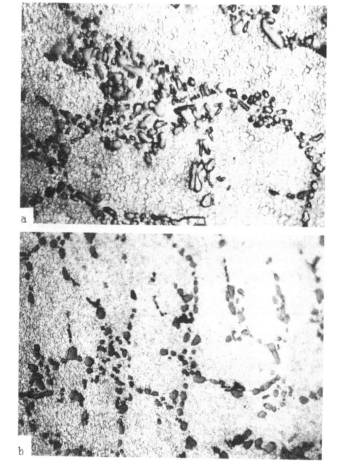

Figure 2.32   Structure of silumin under different casting conditions: (a) $T = T_c = 990$ K; (b) $T = T_c = 1130$ K ($\times 800$).

The specimens were examined in the as-cast condition and after strengthening heat treatment.

In accordance with the results of studies of the cast microstructure it was established that on the series 1 specimens cast at low temperatures, along with primary dendrites of the α-solid solution, large needlelike silicon crystals (called pseudoprimary) occur (Fig. 2.32a). A rise in the cast temperature inhibits their formation and refines the basic structural constituents: the eutectic and the α-solid solution (Fig. 2.32b).

According to the results obtained in the study of the macrostructure of second-series samples, the positive effect of HTT of melt is not inherited. On cooling of the melt after its overheating, silicon "pseudoprimary" crystals appear again. They persist in the structure of castings even after heat treatment.

A quantitative metallographic analysis revealed the changes in such structural characteristics as the mean size of eutectic silicon (L_0), the mean size of the dendritic cell of the

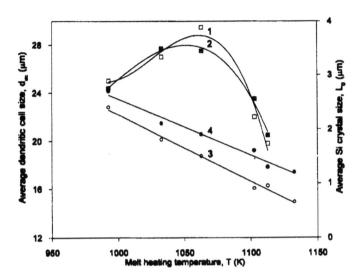

Figure 2.33   Dependence of the average sizes of dendritic cells (1, 2) and crystals of eutectic silicon (3, 4) on the casting conditions for AL9 alloy before (1, 3) and after heat treatment (2, 4).

α-phase (d_{dc}), and the volume fraction of this phase as a function of HTT regimes before the melt solidification.

The sizes of the structural characteristics of cast and heat-treated samples are plotted in Fig. 2.33 as a function of the temperature of the melt heating provided that $T = T_c$. It is evident that d_{dc} and L_c behave in different ways, i.e. an increase in T causes a monotonic decrease in the size of silicon crystals in the eutectic and is responsible for the extremal dependence of linear characteristics of dendrites of the α-phase.

The results of mathematical processing of the numerical values of structural characteristics of all alloy constituents after heat treatment prove that the temperature dependences of the sizes of silicon crystals and dendrites of the α-phase do not differ from the corresponding values in the as-cast state, while the absolute L_0 values increase by 30%.

The measurements of microhardness of dendrites of the α-solid solution in the as-cast condition revealed the growth of H_μ from 580 to 710 MPa with increasing temperature.

The temperature dependences of the ultimate tensile strength σ_u and the relative elongation δ of samples melted according to various regimes of HTT of melt are shown in Figs. 2.34a,b. Clearly, the δ values grow with increasing temperature up to 1100 K, whereas σ_u is less sensitive to changes in the temperature regime of solidification. When using a step-by-step HTT of melt, including the preliminary heating to $T < T_{hom}$ and cooling to $T_c = 990$ K, the increased δ values are not retained. However, these values are twice as high as those measured for castings obtained by the standard technology without HTT.

Thus, cooling of the melt preoverheated to $T \leq T_{hom}$ partly removes the positive effect of raising mechanical characteristics related to HTT.

Alternative results were obtained with the use of technology in which the homogenizing HTT was employed, i.e. at $T > T_{hom}$.

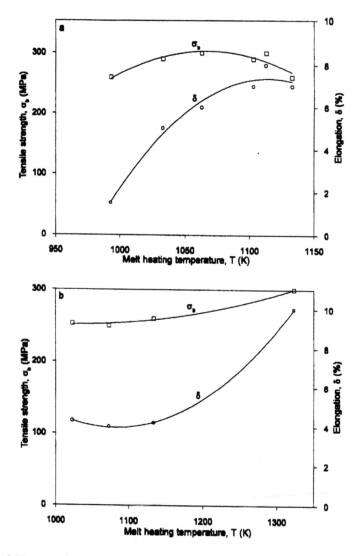

Figure 2.34   Dependence of the ultimate tensile strength and relative elongation in castings of AL9 alloy on the casting conditions: (a) $T = T_c$; (b) T is variable, $T_c = 990$ K.

Figure 2.34b shows mechanical properties of samples cast from liquid silumin heated to 1320 K and then cooled down to $T_c = 990$ K. These properties show that the overheating of the melt above T_{hom} makes it possible to retain high σ_b and δ values even at low cast temperatures.

The results obtained proved the usefulness of HTT of melts as a technological means for improving the properties and quality of materials.

Experimental data on the effect of HTT of melts on the structure and properties of silumins are accounted for by the theory of microheterogeneous structure of Al–Si melts and by specific features of solidification of silumins.

According to experimental data on the properties of liquid hypoeutectic silumins (Section 2.1), microgroups of colloidal sizes inheriting the silicon phase structure are preserved at small overheating above the liquidus temperature. As the temperature increases, we observe refinement and solution of such groups, which ends at temperatures above the homogenization temperature. In the process of solidification of the microhetero- geneous melt, these structures play the role of nucleation centres on which crystals of "pseudoprimary" silicon form. Thus, the observed reduction in the size and number of silicon crystals with the increasing cast temperature of castings can be explained by different microheterogeneous structures of the melt at its heating below T_{hom}. In this context it is clear that interim cooling of such a non-equilibrium melt down to $T > T_L$ promotes the retention in it of centres of crystallization of the silicon phase; as a consequence, stepwise HTT of the melt results in the growth of needlelike "pseudo- primary" silicon crystals responsible for the decrease in plasticity of material.

If the melt is cooled after converting the melt into the homogeneous state (the procedure that was implemented in experiments at 1320 K), the nuclei of the primary silicon phase are absent and the entire process of formation of the solid phase occurs at significant overcooling at the interphase crystal–melt boundary. The latter gives rise, in turn, to the formation of a disperse eutectic structure.

Thus, the decrease in the size of silicon crystals and in the dendritic parameter, which was revealed as the casting temperature increased, as well as the increase in the volume fraction of the α-phase, account for the complex processes of nucleation and kinetics of growth of structural constituents of the alloy controlled by the melt condition specific for each temperature.

The studies described in this section confirmed the interrelation between the methods and technological melting regimes, on the one hand, and the properties of melts and solid materials formed from these melts, on the other hand. They also made possible the determination of optimal HTT regimes in producing castings from the AL9 alloy with plastic properties increased by a factor of 1.5–3.

Chapter 3

Modification as a Means of Controlling the Structure of Liquid and Solid Alloys

3.1 Classification of Modifiers

In scientific literature we find a sufficient number of publications devoted to the theoretical and experimental problems of alloys modification. The most complete and modern description of the nature and mechanisms of alloy modification is presented in Gol'dshtein and Mizin's monograph [1]. The authors present ample experimental material on the inoculation of melts of iron-carbon alloys, the method which consists in the formation of artificial solidification centres. Alternative approaches, as the authors claim, are the formation of barriers at the surface of growing crystals achieved by introducing surface-active additives, as well as various external actions on the solidifying metal: mechanical, electromagnetic, vibrational, and ultraacoustic.

Here we will outline very briefly the main principles of the theory and practice of alloy modification with particular emphasis on the specific features of this phenomenon for aluminium compositions.

Systematic studies of modification processes began as early as 1940s and have been continued by Thiller, Tamman, Hanemann, Danilov, Dankov, Rebinder, and other researchers who assumed the existence of nucleation catalysts, among which a special role was given to various impurities affecting the solidification process. More recently Richards established the basic experimental facts relating to the effect of impurities on the value of melt supercooling, the existence of a particular temperature for impurity deactivation, and the role of alloy overheating. Dankov, and independently Turnbull, developed the crystallographic theory of the effect of nucleation catalysts, whereas Cibula [2], Kuznetsov and Palatnik [3] were the first to point out the need for allowing for the chemical nature of the support. Cibula showed, in particular, that carbides of titanium, vanadium, zirconium, tungsten, and molybdenum serve as effective catalysts of crystal nucleation for aluminium and its alloys. In contrast, chromium and manganese carbides do not exhibit such properties. In connection with this, Totti and Reynolds established that grain refinement occurs if the lattice parameters of a catalyst and solidifying metal do not differ by more than 10%.

According to the classical works of Rebinder and the associated school of thought, it is useful to divide all modifiers into two groups [4]. Group I modifiers create the ultradisperse suspension of individual particles which become solidification centres in the melt. Group II modifiers influence the habitus and kinetics of nucleating crystals due to adsorption on their facets. The mechanism of modification of silumin with sodium was explained by the action of such modifiers.

However, a few works have appeared recently which cast some doubt on the former view of the nature of sodium's modification effect and prove its inoculation mechanism [5, 12].

Of special interest are the results obtained by Esin and Gel'd [1], who showed that the viscosity of liquid silumin grows with Na addition as a result of the formation of solid particles in the melt. Boom [12] identified these particles as a high-melting compound of the ternary sodium silicide type. We can also give another example, which proves the incorrectness of invoking the adsorption hypothesis alone for explaining the modification mechanism. In alloying Al–Mg alloys with various additives it was shown that their surface activity, or inactivity, does not always characterize the modification effect [7, 8]. For example, bismuth decreased most strongly the surface tension in the alloy, but with no modifying effect. The strongest grain refinement of alloys was observed after introducing high-melting additives, such as Zr, Ta, Nb, and B, which attested to the validity of nucleation theory. There are many similar examples of ambiguous treatment of results on modification of aluminium alloys, which means that the classification described above is not uniquely correct.

An alternative well-known classification of modifiers was suggested by Mal'tsev [7]. In his works explicit preference is given to inoculation, with high-melting particles serving as crystallization centres. These particles may be aluminides (Al_3Ti, Al_3Zr), if the modification is accomplished with the aid of additions of transition metals, or borides, which arise directly in the melt in the treatment of aluminium alloys with fluxes (mixtures of salts K_2TiF_6 and KBF_2). Meanwhile, some researchers (for example, Flemings with coworkers) believe that the modification effect in Al–Ti and Al–Zr alloys is related to the peritectic reaction, and primary aluminides of zirconium and titanium serve as crystallization centres [9]. However, it was shown in later works of other authors [13, 14, 18, 24, 37] that modification also occurs at concentrations of additives lower than the peritectic one.

Having generalized the available experimental material, Mal'tsev put forward the idea that the modification mechanism can be specified by purely external signs, for example, by the effect of a modifier on the structure of the material. His classification [7] considers three types of modification processes: the refinement of primary grain-dendrites, the modification of their internal structure, and the change in the eutectic morphology.

Quite different views of the nature of the effect of additives on the material structure were expounded by Samarin [10] and Ershov [11]. The essence of their concept was that the modifying capability of an additive in metals and alloys is determined by its limited solubility in the solid state and by the capability of an additive to stabilize the microheterogeneous structure of the initial melt. The authors develop a model of cluster microheterogeneity of liquid metals and assess the effect of additives on the process of crystal nucleation by their influence on the size and stability of clusters in melts prior to the onset of solidification.

The correctness of this concept was corroborated in [10, 11, 13–25]. The authors of [10, 11] recognize the division of modifiers into two types (after Rebinder) but define them in a different way [11]. In particular, according to their terminology, type I modifiers are particles serving as crystallization centres and representing nuclei at the surface of which individual clusters are grouped. Such a combination of a nucleus and a surrounding cluster envelope (polycluster) must be thermodynamically stable as a whole, both at the solidification temperature and above it. By setting some criteria which polyclusters should satisfy, the authors determined conditions that characterize the efficiency of type I modifiers. All the calculations were made for iron-based alloys.

The choice of effective type II modifiers was based on experimental results showing that the best modifiers are additives with more distinctly pronounced metallic properties than the matrix. From this point of view, elements that easily give up valence electrons and form metallic bonds should facilitate the strengthening of intercluster bonds and their merging, whereas additives forming non-metallic-type bonds (accepting valence electrons), favour the separation of clusters by forming a barrier that fixes the cluster and hinders its growth. As a factor specifying the relative modifying capability, the authors suggest the inter-relation between the solubility of impurities in the cluster volume and the effective ionization potential commensurate to electronegativity. According to the developed criteria, the "strongest" type II modifiers for steels are Ba, Ca, and Sr.

In spite of the fundamental approach to the choice of suitable modifiers, such an approach has disadvantages because the limited set of its criteria does not take into consideration the great diversity of actual factors influencing the process of modification in steels (for example, the crystal–structure factor). The role of the latter was proved convincingly by Dankov [6] and Palatnik [3].

In conclusion, we present an alternative point of view of the alloy structure modification mechanism developed by Chernov and Busol [19], who criticized both the absorption and inoculation concepts. According to their conclusions, the modifying action of additives soluble in the melt (in particular, additives with the distribution coefficient $K_0 < 1$) can be explained by the fact that in view of the different solubility in the solid and liquid states such additives are driven off by the crystallization front into the melt. These arguments rely upon the theory of concentration supercooling developed by Tiller, Millens, and Sekerka. The authors of [19] suggest that in some cases concentration supercooling may be sufficient for independent nucleation of new crystallization centres, which is similar to the action of inoculants. Moreover, these authors believe that the pushing out of some impurities by the crystallization front exerts a barrierlike, decelerating action on the crystal growth which results in the formation of a fine-grained structure.

This is by no means a complete analysis of the theoretical and experimental data on the mechanisms of modification of metals and alloys available in our scientific literature. This analysis gives evidence of ambiguous treatments and incomplete ideas in this important field of materials science. Further, in the discussion in Chapters 3 and 4 of our original results on the modification of aluminium alloys, we will examine the possibility of implementing other mechanisms of this interesting phenomenon [13, 14, 20–24].

3.2   Modifying Master Alloys Prepared with the Use of HTT of the Melt

3.2.1. Methods for production and structure of master alloys

Master alloys are used to alloy aluminium alloys with various, most frequently high-melting components. Of particular importance are the alloys of aluminium with transition metals: in the last few decades a new class of aluminium-based structural materials has been developed in which transition metals are present as additives [26–28].

A great body of experimental data shows that the addition of transition metals to aluminium alloys changes the size of cast grains and causes the formation of anomalously supersaturated solid solutions [7, 26–28]. For example, zirconium additives reduce the tendency of alloys to stress corrosion by affecting the degree of recrystallization, and raise both the alloy strength, due to alloying the aluminium-based α-solid solution, and the plasticity of materials under the influence of the structure modification. As follows from Section 3.1, transition metals – scandium, titanium, and zirconium – are the most effective structure modifiers.

A variety of new methods of special treatment and producing charge and master alloys have been developed in recent years with the aim of improving the quality of castings and ingots. These methods are based on the concept of existence of hereditary linkage in the charge–melt–castings system which was discussed in Section 1.6 [13, 17, 23–25].

Whilst an extensive practice in using various master alloys, the search for new methods of their production providing the enhanced modifying capability of alloys remains a pressing problem.

In this section we describe the original methods of producing Al–Zr, Al–Ti, and Al–Sc master alloys based on specific structural features of their melts and generalize experimental data on the role of the master alloys structure in forming the cast structure of alloys [13, 23].

Prerequisites for the development of new technologies for producing master alloys were the results obtained on model alloys of aluminium with transition metals [20–22]. It was established that the use of HTT of melts in combination with high cooling rates allows one to vary over a wide range both the size and the form of growth of structural constituents of alloys. Of obvious practical interest was an investigation of the role of HTT of a melt in producing master alloys by means of the standard casting technology, i.e. at low cooling rates close to those achieved in permanent mould casting.

As objects for investigation we chose the commonly employed master alloys of Al–2% Zr, Al–3.8% Ti, and Al–2% Sc compositions obtained preliminarily by means of the standard industrial pig-cast technology [41].

The methods for producing master alloys were as follows: charge materials with mass of about 20 g were placed in alundum crucibles and heated in a chamber furnace with a a silit heater to the experimentally preselected temperature (usually in the range 1220–1520 K). The melt was soaked at this temperature for 30 min and solidified in the form of ingots (at the cooling rate $V = 10$ K/s during solidification) or in the form of ribbons of 200 μm thick (at $V = 10^4$–10^5 K/s). For brevity, we use the following notations of different master alloys: SMA is the master alloy obtained with standard technology; HMA is the master alloy

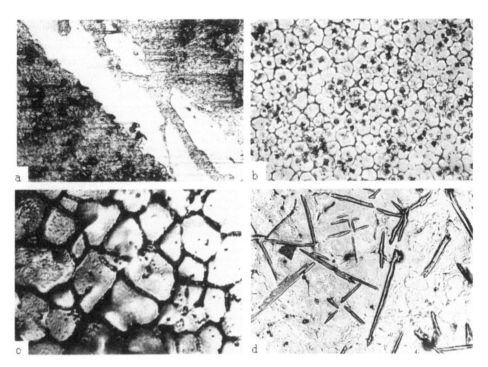

Figure 3.1   Master alloys obtained with the use of experimental (a–c) and standard (d) technologies: (a) Al–2% Zr OMA ($\times 1000$); (b) Al–2% Zr HMA ($\times 1000$); (c) Al–3.8% Ti HMA ($\times 2000$); (d) Al–2% Zr SMA ($\times 200$).

obtained by rapid solidification after HTT of the melt; and OMA is the master alloy obtained with normal casting rates from the preoverheated melt.

When studying the structure of ingots made of master alloys, the following regularities were revealed: irrespective of the initial temperature of melt heating, crystals of intermetallic compounds Al_3Zr and Al_3Ti with DO_{23}- and DO_{22}-type structures, respectively, are present in the ingot structure; the sizes of aluminides, depending slightly on the HTT of the melt, lie within the limits of 100 μm; heating the melt to temperatures above 1450 K causes the appearance of defects in the structure of aluminide crystals in the form of subboundaries and cracks.

The last effect is related to the fact that an increase in the temperature of melt heating enhances the formation and development of non-equilibrium forms of crystal growth and leads to dendrite formation. As a result, roughness and projections arise, which reflect the morphological instability of the interphase boundary. The presence of projections and discontinuities leads to the capture of matrix melt and its subsequent solidification inside the crystal (this is illustrated clearly by the presence of bright regions in Fig. 3.1a), which also increases the number of defects.

The defect structure of aluminide crystals influences their microhardness. In our experiments with defect-free aluminide crystals we obtained H_μ = 5500 MPa and 5900 MPa, which agrees reasonably well with the experimental data [30] for Al_3Zr and Al_3Ti crystals, respectively. For defect-containing crystals with defects present in the structure of master alloys obtained from the overheated melt, the H_μ values are lower by a factor of 1.5–1.7.

According to [31], the thermal stability of phases precipitated in solidification and having the same melting temperature is proportional to their hardness. As a consequence, crystals with defects are less stable and dissolve more efficiently in the matrix melt. As will be shown below, this property plays an important part in estimating the modifying capability of master alloys.

When developing a new technique for producing master alloys by the method of rapid solidification, the goal was to obtain a master alloy with maximum number and minimum possible size of intermetallic compounds.

Below we present the results that were used to determine the optimal HTT regimes for obtaining the Al–2% Zr master alloy by a rapid quenching technique with $V = 10^4$ K/s.

Table 3.1   Number (N) and sizes (L_0) of Al_3Zr crystals as a function of HTT of the melt in the Al–2% Zr alloy.

T, K	ΔT, K	L_0, μm	N, mm^{-2}
1170	100	16	1×10^4
1270	200	10	1×10^4
1370	300	6	2×10^4
1420–1470	350	2	4×10^4

At higher heating temperatures ($\Delta T \geq 450$ K), the primary solidification of aluminides is suppressed and an anomalously supersaturated solid solution growth occurs.

Thus, we have the following optimal conditions for obtaining an Al–Zr alloy satisfying the requirements formulated above: $V = 10^4$ K/s and $\Delta T = 350$ K (which corresponds to $T = 1420$ K). The structure of such a material consists of aluminides of metastable phases of size no more than 5 μm, which are uniformly distributed over the cross-section of the alloy (Fig. 3.1b).

In a similar way the conditions for obtaining Al–Ti and Al–Sc master alloys were determined. It was shown that the region of the most stable growth of crystals of metastable modification is displaced to the region of higher cooling rates and becomes equal to 10^5 K/s at $1370 < T < 1520$ K. The typical structure of this alloy is given in Fig. 3.1c (the mean size of crystals is 3 μm).

As was shown experimentally, the solidification parameters providing the formation of a fine-grained structure in the Al–2% Sc master alloy remain invariable ($V = 10^5$ K/s, $\Delta T > 400$ K). The structure of this alloy is shown in Fig. 3.1c.

Thus, summing up all the experimental data listed above, we can determine the solidification conditions leading to the formation of master alloys with the given structure: it is necessary to perform high-temperature preheating of the melts ($\Delta T = 350$–400 K) with subsequent solidification in the range of cooling rates from 10^4 to 10^5 K/s.

Taking into account the common features in the structure of aluminium melts with transition metals and the hereditary linkage between the structures of alloys in the liquid and solid states, the authors of the works [20–23] showed that the optimal regimes of HTT of melts for obtaining master alloys with special modifying properties can be inferred from the temperature dependences of their structure-sensitive properties. It was found that the temperature of overheating the melt above the liquidus should not exceed the temperature T_{hom} of its transition from the microheterogeneous state into the true solution state but should be close to the latter (Table 3.2).

The presence of a fine-grained structure with uniform distribution of the intermetallic phase in rapidly cooled alloys is proof of the fact that aluminium melts with transition metals retain their microheterogeneous structure up to high temperatures, i.e. to T_{hom}.

Table 3.2   Optimal regimes of HTT of melts of master alloys.

Alloy composition	HTT of melt	T_{hom}, K
Al–2% Zr	1490–1500	1520
Al–2% Sc	1570–1600	1670
Al–3.8% Ti	1270–1320	1510

The use of special HTT of liquid metals in the production of aluminium master alloys with transition metals has led to the development of alloys differing in structure from previously known alloys.

Let us consider the role of master alloys in the modification of aluminium and aluminium-based alloys.

3.2.2. Properties of master alloys

To assess the modifying properties of master alloys we carried out special experiments with the aim of carrying out the following tasks:
   • to verify the effect of the size of intermetallic phases on the modifying capability of alloys;
   • to reveal additional structural features enhancing the modifying effect of master alloys (such as the defect structure of aluminides and the type of their crystal lattice);
   • to establish the optimal conditions for introducing various master alloys and the area of their application in the production of aluminium alloys.

The modifying capabilities of master alloys can be assessed primarily by way of solidification of the α-solid solution which forms the base of such alloys.

With this purpose, we estimated by quantitative metallographic analysis the cast grain size in Al–2% Zr and Al–3.8% Ti master alloys obtained with the use of different technologies: the standard technology with casting into pigs (Fig. 3.1d) and experimental technology – pouring out the overheated melt onto the rotating copper cone (Figs. 3.1b,c).

As follows from the structures shown in Figs. 3.1b,c,d, in the first case we obtain a coarse dendritic structure with grains more than 150 μm in size, which is morphologically unrelated to the solidifying first intermetallic phases. In the second case, a modified structure with non-dendritic grains is observed in the master alloy; the grain size does not exceed 10 μm and becomes comparable with the size of dendritic cells. According to terminology accepted in modern materials science, such a structure is called subdendritic [28, 29].

Thus, primary solidification of aluminides affects the formation of the basic structural constituent of the master alloy: grains of the α-solid solution. In the case when metastable disperse aluminides precipitate first in solidification, a uniform disperse crystalline structure forms, which is similar to that obtained by special melt processing, for example, in ultrasonic treatment (this phenomenon is discussed in greater detail in Chapter 4).

Below we present experimental data on the relative efficiency of modification of the structure of commercial aluminium and its alloys by small Zr additions introduced with the aid of master alloys of the same composition obtained with various technologies [13].

Table 3.3 contains the thermal physical parameters of solidification of such master alloys of the Al–2% Zr composition.

Table 3.3   Conditions for preparation of Al–2% Zr master alloys.

Alloy composition	Master alloy type	V, K/s	T_c, K
Al–2% Zr	SMA	10	1370
Al–2% Zr	HMA	10^4	1520
Al–2% Zr	OMA	10^0	1520

Table 3.4   Regimes of ingot casting with the use of Zr-alloyed commercial aluminium.

Zr content	$T = T_c$, K	τ, min	Master alloy type	Note
0.1	1070	30	SMA	
0.15	1070	30	SMA	Curve ——
0.2	1070	30	SMA	Fig. 3.2
0.2	970	30	SMA	
0.1	1070	30	HMA	Curve ·····
				Fig. 3.2
0.15	1070	30	HMA	
0.2	1070	30	HMA	
0.1	1070	15	HMA	
0.15	1070	15	HMA	
0.2	1070	15	HMA	
0.1	970	15	HMA	Curve ---
				Fig. 3.2
0.15	970	15	HMA	
0.2	970	15	HMA	

Alloys with various zirconium additives (0.1, 0.15, and 0.2%) were prepared from commercial aluminium and Al–Zr master alloys with a a silit heater. The temperature–time characteristics of the introduction of master alloys in the base melt, the cooling rate, and the cast temperature were preassigned in advance. The regimes of ingot casting from commercial aluminium alloyed with zirconium are listed in Table 3.4, and the results of metallographic examination are shown in Figs. 3.2 and 3.3.

It was established that the modification depends on the master alloy type and temperature conditions of its introduction in the melt. When SMA was used to alloy the material with

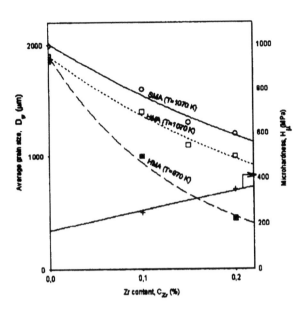

Figure 3.2 Dependence of average grain size and microhardness of commercial aluminium ingots on zirconium concentration for different master alloys: SMA (1); HMA (2, 3).

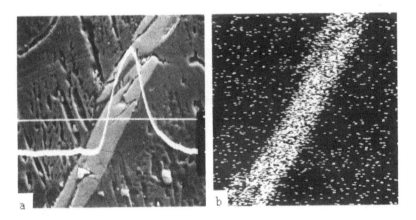

Figure 3.3 Electron probe microanalysis of intermetallic compounds in the structure of the ingot alloyed with zirconium with the use of SMA; (a) secondary electron image ($\times 1000$); (b) zirconium K_α image.

zirconium, excess crystals of the Al_3Zr intermetallic compounds with a plate structure similar to that of intermetallics of the master alloy formed in the structure of ingots, independently of their composition, against the background of the dendritic structure with widely different grain sizes (Fig. 3.3a). The phase identification was performed using the EPMA technique (Fig. 3.3b). The presence of such crystals in the structure of cast material is due to the limited solubility of zirconium in the α-solid solution and was noticed by many authors [7, 26, 27].

Ingots alloyed with zirconium with the use of HMA possess a uniform structure and have grains of rounded shape. Dendritic growth in them is suppressed. As the temperature of introduction of this master alloy into the base melt increases from 970 to 1070 K, its modifying effect decreases and grains acquire a dendritic structure.

According to metallographic, X-ray diffraction, and electron probe microanalyses, the distinctive feature of the structure of all ingots alloyed with the aid of HMA is the lack of primary Al_3Zr intermetallic compounds. This finding indicates that an anomalously supersaturated (with 0.2% Zr) α-solid solution is obtained. This inference is also confirmed by the increase (from 250 to 370 MPa) of microhardness values after Zr alloying.

In the next series of experiments we tested the technology of producing castings from the aluminium melt subjected to preliminary heat treatment at 1170–1520 K. Zirconium (from 0.2 to 0.4%) was also added to liquid aluminium with the aid of three different master alloys of the same composition (Al–2% Zr) differing in preparation technologies (Table 3.3).

The regimes of HTT of melts before the melt solidification and characteristics of master alloys are presented in Table 3.5.

The concentration dependences of linear grain sizes in castings are plotted in Fig. 3.4 to estimate comparatively the results of metallographic studies. One-phase structures of the solid solution free of excess aluminides are shown in this figure by the hatched symbols (*).

When combined with modification, the HTT of the melt affects substantially the structural characteristics of the material. Let us consider this effect in greater detail. If the melt is preoverheated to 1170 K, the following results are obtained: for the alloy with 0.2% Zr the one-phase structure with finest grains is observed in castings obtained with OMA. The use of HMA also results in the formation of a structure without precipitates of excess zirconium aluminides but its modifying properties are worse.

To verify the data of metallographic analysis we used a specially devised method of resistivity measurements [13, 40], which confirmed that the use of experimental master alloys makes it possible to reliably obtain anomalously Zr-supersaturated α-solid solutions, even at standard rates of cooling castings.

Table 3.5 Casting regimes and characteristics of the cast structure of ingots obtained from zirconium alloyed aluminium.

Zr concentration, %	$T = T_c$, K	Master alloy type
0.2	1170	SMA
0.2	1170	HMA
0.2	1170	OMA
0.2	1520	SMA
0.2	1520	HMA
0.2	1520	OMA
0.4	1170	SMA
0.4	1170	HMA
0.4	1170	OMA
0.4	1520	SMA
0.4	1520	HMA
0.4	1520	OMA

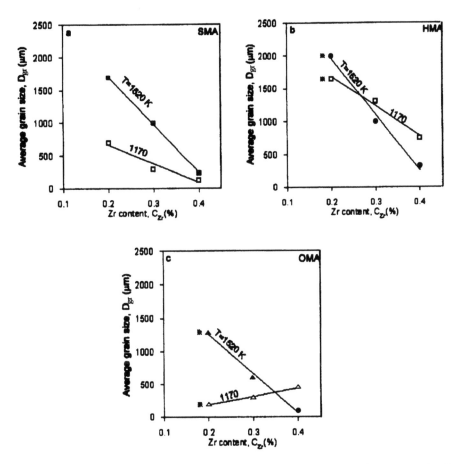

Figure 3.4   Concentration dependences of the mean size of cast grains in ingots made of commercial aluminium alloyed with zirconium with the use of various master alloys; (a) SMA, (b) HMA, (c) OMA.

In the case of using SMA, we obtained castings of a two-phase constitution; the amount of excess aluminides in the alloy is retained even as the melt heating temperature increases up to 1520 K. As was found, high-temperature treatment of the melt also decreases the modifying properties of OMA and coarsens cast grains in the alloyed material.

This result seems at first glance trivial because the coarsening of grains with melt overheating is accounted for by impurity deactivation (see details in Chapter 4). With allowance for the heterogeneous mechanism of solidification of materials of commercial purity, the significance of this phenomenon in the structure formation must be acknowledged. However, for aluminium alloyed with transition metals, the role of over-heating is much more complicated. We will attempt to demonstrate by our examples that situations are possible in which overheating has the opposite effect, i.e. decreases the grain size. Such a situation is observed when the zirconium content in aluminium increases to 0.4% (alloying with the use of OMA and HMA). In this case we fail to suppress completely the primary solidification of aluminides. However, their amount in ingots is lower by a factor of two while the zirconium content in the α-solid solution exceeds significantly the equilibrium concentration and reaches 0.21–0.23%.

Thus, the application of HTT of melts in combination with the use of special master alloys enables one to increase the zirconium content in the α-solid solution up to 0.20–0.23 wt. % and improve significantly, after subsequent heat treatment in the solid state, the strength properties of the material.

On the basis of the above experimental data, it was demonstrated convincingly that the structure of master alloys plays an important role in forming the final structure of alloys with additives. The best properties of master alloys obtained by rapid solidification may be determined by the following characteristics: the highest bulk density of aluminides, their small size, and crystal lattice conjugated with aluminium. While the positive role of the first two factors may be thought of as proved, the influence of the type of crystal lattice of aluminides on the modifying capability of master alloys does not follow with certainty from the above results. To find out the role of this factor, we performed special studies in which a comparison of HMA with a master alloy having the same chemical composition and size of aluminides was made. Such a fine crystalline MA was obtained in the form of a ribbon by pouring the melt into rotating water-cooled rollers [25]. In such an alloy a two-phase structure of the α-solid solution forms with primary Al_3Zr aluminides of 20 μm in size, having a tetragonal crystal lattice of the DO_{23} type.

A comparison of modifying capabilities of the given and experimental HMA was made for commercial aluminium alloyed with zirconium in amounts of 0.1–0.2%. The introduction of master alloys was brought about with the use of identical technologies involving heating to 970–980 K, 15 min. holding, and solidification into 15–20 g ingots at a cooling rate of 0.5 K/s. Such conditions were chosen experimentally and proved to be optimal for master alloys of the given type. The results of metallographic examination are summarized in Table 3.6.

Table 3.6 Alloying conditions, composition, and structural characteristics of ingots prepared from commercial aluminium.

Master alloy type	Zr content, %	Grain size, μm	Presence of excess phase
MA	0.1	1800	no
	0.15	1600	yes
	0.2	1200	yes
HMA	0.1	1000	no
	0.15	550	no
	0.2	400	no
	0	2000	no

These experimental data show once again that the effect of cast grain refinement is observed at low concentrations and that the conjugation of lattices of Al_3Zr crystals and matrix significantly contributes to modification.

The following points can be distinguished from the experimental material listed above:

• The master alloys (HMA, OMA) developed with the use of original methods combine well modifying and alloying properties;

• The efficiency of using one or another property is determined by the regime of HTT of the melt;

• At low-temperature and short-term treatment of melts, it is appropriate to use HMA as a modifying master alloy providing good structure refinement with ensuing positive properties, the most important of which is the enhanced plasticity of the material;

• The use of high-temperature treatment of the melt in combination with alloying with transition metals permits one to increase significantly (by a factor of 1.7–2.9) the impurity content in the base alloy and opens wide possibilities for increasing the strength characteristics of the material.

3.3   Role of Low-melting Surface-active Additives in the Formation of the Structure of Aluminium Alloys

3.3.1. Structure modification

From the special literature devoted to modification of aluminium alloys we know well the works [4, 32] the authors of which showed for the first time that the structure of metals and alloys in their solidification can be changed by introducing surface-active elements. These works constituted subsequently the basis for studies in the theory and practice of modification.

The mechanism of cast grain refinement proposed by these authors is based on the slowing down of the rate of growth of the solid phase by means of creating a "barrier" between the crystal and the melt caused by adsorption of low-melting surface-active additives at crystal facets.

Upon considering the modern works based on the concept of the structural heterogeneity of melts [15, 24, 33, 34], it becomes evident that the effect of small additions is much more complicated and requires further investigations.

In this section we present the experimental results on the influence of low-melting micro-additions on the structure of aluminium alloys in the liquid and solid states; part of these data was published in [14, 24, 35].

Initially, the well-studied Al–Si system was chosen for investigation, with tin and zinc as alloying additives. The base Al–26% Si alloy was prepared from A999 grade aluminium and silicon of semiconductor purity. Zinc was introduced with the aid of an Al–6% Zn master alloy and tin was introduced in the melt directly before solidification. Samples in the form of ribbons were obtained with the use of a high-speed solidification installation at $V = 10^2 - 10^4$ K/s. The procedure included the use of HTT of melts, with the melt heating temperatures varying in the range 1170–1470 K.

In Chapter 2 (Section 2.2) we described in detail the morphological and size changes in the forms of growth of crystals of primary and eutectic silicon in Al–26% Si alloys, which were caused by the effect of HTT and rapid solidification. A knowledge of these changes and a comparison with the results described in this chapter makes it possible to find out the role of small additions in the structure formation in the process of solidification of silumins.

The microstructure of the Al–26% Si–0.1% Sn alloys is presented in Fig. 3.5. Clearly, the main structural constituents of the alloy with added tin are retained but some new

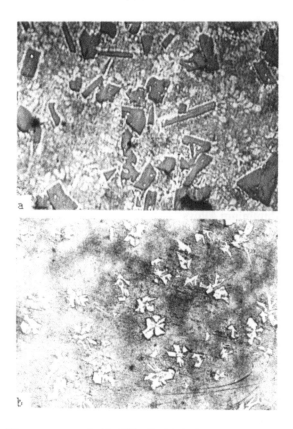

Figure 3.5   Microstructure of Al–26% Si–0.1% Sn alloy under different casting conditions; (a) $V = 10^3$ K/s, $T = 1370$ K; (b) $V = 10^4$ K/s, $T = 1470$ K (×200).

growth features appear. For example, the growth of primary faceted silicon crystals becomes more stable and holds up to 1370 K (Fig. 3.5a). Moreover, large anisotropic dendrites – the predominant form of growth in solidification of the overheated Al–26% Si alloy – break up after alloying with tin and become more equiaxed. This process is more pronounced in the structure of the Al–26% Si–0.1% Sn alloy which is cast with a cooling rate of 10^4 K/s; the maximum refinement and transition to spherulite-like crystals is attained when the cast temperature rises to 1470 K (Fig. 3.5b).

An increase in the tin content to 0.2% promotes further modification of the alloy structure and the flattening of morphological and size features caused by HTT of the melt. In all the alloys after introduction of tin additives the amount of the α-phase near primary silicon crystals decreases, which makes the structure more uniform.

As follows from the quantitative metallographic data, the size of crystals declines with increasing temperature, independently of their morphology; the volume fraction of primary silicon precipitates changes with temperature non-monotonically but the tendency to its reduction is retained irrespective of cooling rate (Figs. 3.6a,b).

A more detailed analysis of the temperature curves plotted in Fig. 3.6b shows that for each composition a definite melt overheating temperature exists at which a sharp increase

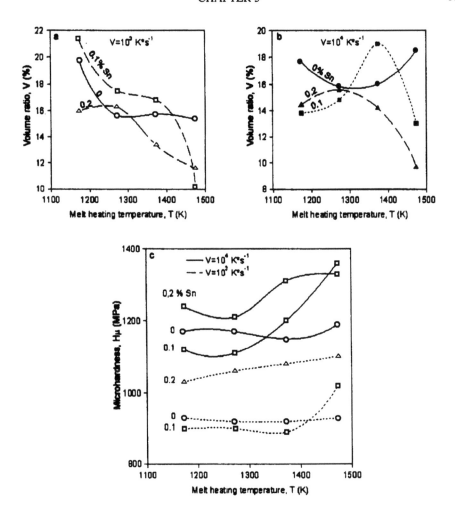

Figure 3.6 Variation with temperature of the volume fraction of primary silicon crystals (a, b) and of the microhardness of the eutectic (c) in Al–26% Si alloys with tin obtained at different cooling rates: (a) 10^3 K/s; (b) 10^4 K/s.

in the volume fraction of the silicon phase occurs in the ribbon structure. The higher the tin content in the alloy, the lower is this temperature. The cause of this phenomenon is discussed in Section 3.4.

The indicated changes in the morphology and size of the primary phase influence, no doubt, the conditions of eutectic formation, which affect in turn the structure of this phase, relative phase proportions, and microhardness.

As a result of the studies conducted in this work we established that the basic modifying effect of tin additives manifests itself in the change of the eutectic type due to the refinement and transformation of forms of growth of eutectic silicon crystals. In particular, at $T = 1170$ K and $V = 1000$ K/s in the alloy containing 0.2% Sn we observe the formation of a mixed eutectic consisting chiefly of thin needles of the silicon phase (Fig. 3.7a),

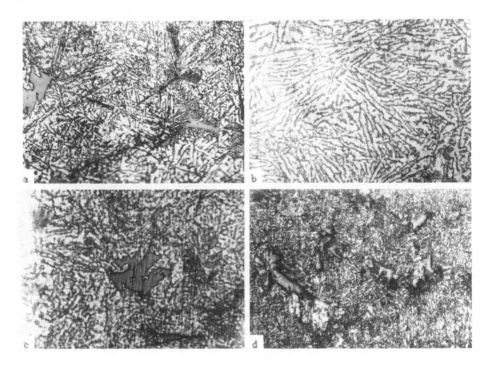

Figure 3.7   Eutectic in Al–26% Si alloy with tin (a, b) and zinc (c, d) at $T = 1170$ K (a, c) and $T = 1470$ K (b, d) ($\times 500$).

whereas for the alloy without tin obtained under identical conditions the most typical is a coarse plate eutectic.

As follows from the microhardness data (Fig. 3.6c) and quantitative metallographic analysis, the introduction of 0.1% Sn begins to exert a modifying influence and increases H_μ only after high-temperature heating of the melt before its solidification ($T > 1370$ K). If the alloy contains 0.2% Sn, the absolute values of eutectic microhardness at all temperatures of heating of liquid metal are higher than for the alloy with 0.1% Sn, and the dependence of H_μ on T is pronounced more slightly. The latter points to the more effective influence of microalloying on the process of structure transformation as compared to HTT of the melt. Moreover, at 1470 K the amount of eutectics increases significantly and the alloy solidifies like the quasieutectic one (Fig. 3.7b).

The effect of zinc microadditions (0.1 and 0.2%) on the structure of hypereutectic silumin was examined in a similar manner. The structure of an Al–26% Si–0.1% Zn alloy obtained by rapid quenching from various melt temperatures is shown in Figs. 3.7c,d. As T increases, the sizes of crystals of primary silicon decrease and a fine-grained structure forms with more uniform distribution of primary phases. The results of the quantitative metallographic analysis are presented in Fig. 3.8a. As follows from the data obtained, the effect of abrupt grain refinement in alloys with zinc additive is observed after heating liquid metal to 320–1470 K. The use of such HTT technology made it possible to obtain 10 μm silicon crystals in ribbons cast with $V = 10^3$ K/s, the result that has been achieved in alloys without additives at an order of magnitude higher cooling rates.

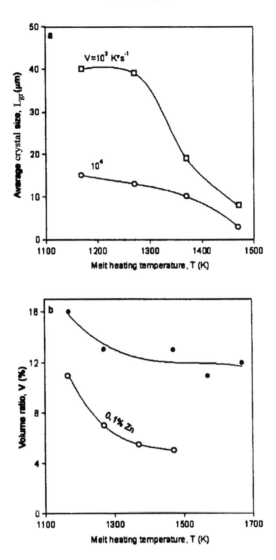

Figure 3.8   Temperature dependences of the mean size (a) and the volume fraction of primary silicon crystals (b) in Al–26% Si alloys with tin obtained at different cooling rates.

The modifying effect in silumin alloyed with zinc also manifests itself in the quantitative proportion of phases formed after solidification. Similarly to the results of experiments on alloying with tin described above, the suppression of primary solidification of silicon in ribbons cast at $V = 10^4$ K/s is observed after overheating the melt to specific temperatures ($T \leq 1470$ K). This is clearly illustrated by Fig. 3.8b. For comparison, we point out that the similar structure in binary silumin is observed in cast ribbons obtained at $V = 10^5$ K/s from a melt preoverheated to $T \leq 1720$ K. In alloys with 0.2% Zn, the tendency for a decrease in the linear sizes of silicon manifests itself much weaker, but the eutectic morphology changes significantly: the eutectic becomes more globular and fine-grained.

Thus, the combined effect on the melt of HTT and microalloying with low-melting additives changes considerably the as-cast structure and facilitates the formation of a quasi-eutectic structure in an alloy of hypereutectic composition at lower cooling rates and lower casting temperatures.

In connection with the growing demand for an increase in heat-resistant properties of cast alloys, they are alloyed with nickel [28]. The combination of high mechanical characteristics with good casting properties of such alloys depends directly on the structure of castings. Unfortunately, existing technologies do not always provide the desired quality of articles. This is additional stimulus for developing new effective methods of improving cast structure by microalloying with various additives.

We investigated the influence of low-melting additives (antimonium, zinc, magnesium, bismuth, tin) on the structure of an Al–26% Si–6% Ni alloy. The specimens were obtained under identical solidification conditions by casting in a permanent copper mould with a cooling rate of 100 K/s at 1170 K/s. Additives were introduced directly into the Al–26% Si melt before casting. The content of microadditions is indicated in Table 3.7.

Precision quantitative local metallographic analysis performed on a computerized materials-science installation "SIAM-340"showed significant changes after alloying in all structural constituents of the alloy: primary crystals of silicon and nickel aluminide, as well as a complex ternary eutectic (α + Si + Al_3Ni). Generalizing ample experimental material, we can note the following regularities:

• the morphology and sizes of primary silicon crystals in the ternary alloy are influenced by the same additives as in the binary alloy, i.e. zinc and tin;

• additives of antimonium, magnesium, and bismuth change mainly the quantity and sizes of crystals of primary nickel aluminides, as well as mutual proportions and morphology of eutectic phases. A fine-grained modified eutectic forms in all alloyed samples; its micro-hardness varies from 1250 to 1450 MPa;

• introduction of microadditions of tin and zinc decreases silicon segregation across the casting cross-section and promotes creation of a more uniform and fine-grained structure.

Thus, the material described above indicates that microalloying with low-melting additives in conjunction with HTT of melts is an effective tool for modifying the structure of some casting alloys. Experiments on modification of hypereutectic silumins with tin or zinc revealed an interesting feature of this phenomenon: the effect of refining the structure in an alloyed ternary alloy shows up at lower casting temperature than in a binary alloy.

3.3.2. Controlling forms of growth of crystals of various structural constituents of alloys

As was shown in the preceding section, microadditions of zinc and tin influence not only the amount and dimensions of silicon crystals, but also their structure enhancing stability of faceted forms of growth.

It is known that the growth rate of crystals depends on their morphology and is determined by various growth mechanisms, with faceted crystals growing faster than rounded (see Section 2.2). This factor can also be used to modify the structure because the slowing down of the growth rate at the same number of crystallization centres facilitates the formation of finer phases.

Table 3.7   Concentration of microadditions in the Al–26% Si–6% Ni alloy.

Additive	Concentration, wt. %	Concentration, at. %
Zn	0.2	0.07
Sn	0.2	0.05
Bi	0.1	0.01
Sb	0.05	0.01
Mg	0.5	0.5

Zinc and tin are controlled impurities in aluminium alloys and their content varies over a wide range depending on aluminium grade. For example, in high-purity aluminium (grade A999) the content of zinc and tin does not exceed 10^{-5}%, while in aluminium of commercial purity (grade A8) [Zn] < 0.04% and [Sn] < 0.02%.

It was of practical interest to study the effect of these impurities on the alloy structure, in particular, on the morphology and sizes of crystals of primary silicon and eutectic. With this purpose we conducted a special series of experiments in which alloys of the same chemical composition (Al–17% Si) but with the base metal differing in purity (A999 and A8) were melted in a vacuum furnace. Next, using the technology described in Section 2.2, samples were prepared at various parameters of high speed solidification. The range of heating temperatures of liquid alloys varied from 827 to 1470 K, and the cooling rate ranged from 10 to 10^4 K/s.

For clarity and comparison of the results obtained, we constructed schemes of morphological stability and quantity of crystals of primary silicon depending on casting conditions of alloys based on aluminium of special (Fig. 2.16) and commercial purity (Fig. 3.9a). As follows from Fig. 2.16a, in Al A999-based alloys the dendritic form is the dominant form of growth of primary silicon crystals in a wide range of solidification conditions. The transition to the faceted form of growth is observed in solidification at $V < 100$ K/s and $T < 1220$ K. In Al A8-based alloys, primary silicon grows only in the form of polyhedral faceted crystals. An increase in V and T narrows the formation zone and leads to the decrease in linear size of the silicon phase in both alloys. Refinement of silicon crystals is accompanied by the decrease in the amount of silicon in the alloy volume, which is evidenced by the curves plotted in Fig. 3.9b.

All changes in the form and kinetics of crystal growth under the above conditions give rise to a nearly complete suppression of primary solidification of crystals and create prerequisites for nucleation of the quasieutectic in accordance with the mechanism described in Chapter 2.

We determined the minimum cooling rates at which a complete suppression of silicon primary solidification is observed and studied the influence of the melt overheating on this phenomenon. Comparing these experimental data with the results obtained by other authors [36], we may indicate that the minimum cooling rate values fall with decreasing purity of the base material from 10^5 to 10^4 K/s. With regard to the effect of HTT of melts on the character of solidification of primary phases, the experiment described above shows the additional possibility of flexibly controlling the form of growth and the number of phases and, accordingly, the properties of the material.

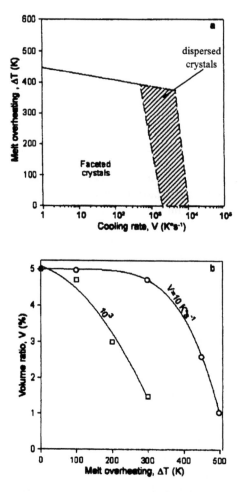

Figure 3.9 Zones of growth of primary silicon crystals (a) and their number (b) in
Al–17% Si alloy (aluminium base metal of commercial purity) at different solidification
parameters.

Comparing morphological features of growth of the Al–Si eutectic in alloys of the same
composition but with base material of a different purity (see Figs. 2.22 and 3.10a), we may
conclude that melt solidification on the basis of commercial aluminium occurs under more
equilibrium conditions and at lower supercooling. For example, a modified eutectics forms
in the structure of samples obtained at $V = 10^3$ K/s (for alloys based on aluminium of special
purity $V = 10^4$ K/s), whereas a coarse plate eutectic is completely absent. The transition
from the needlelike eutectic to the modified globular eutectic is accompanied by a drastic
refinement of its structural constituents; in particular, the size of crystals of eutectic silicon
does not exceed 1 μm and the microhardness of such a structure is much higher than the
microhardness of an unmodified structure (~1200 MPa, Fig. 3.10b).

The effect of microadditions present in alloys of commercial purity on the forms of
growth of crystals was revealed also in the solidification of aluminides, for example, Al_3Zr

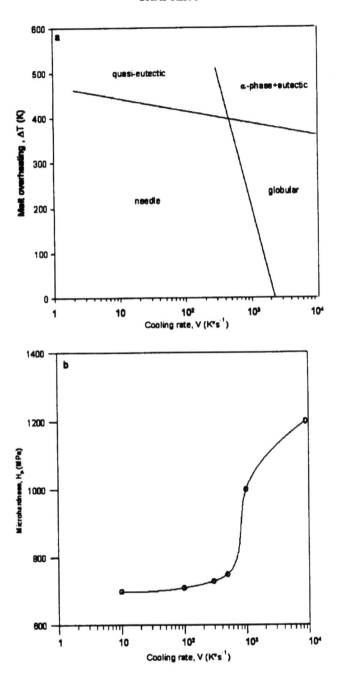

Figure 3.10 Morphology of eutectic (a) and microhardness (b) in Al–17% Si alloy (aluminium base metal of commercial purity) as a function of solidification parameters.

crystals in the alloy of the hyperperitectic Al–2% Zr composition [20]. We compared the structures of two alloys of the same chemical composition but melted under identical

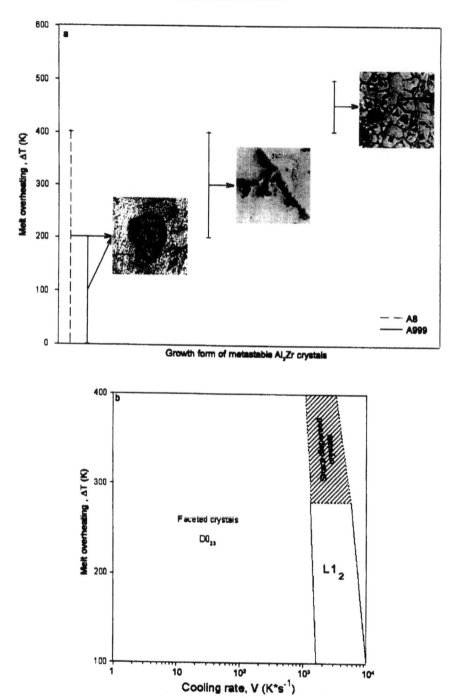

Figure 3.11   (a) Dependence of forms of growth of zirconium aluminides (the $L1_2$ structure type) on ΔT and purity of base material and (b) the regions of stable growth in Al–2% Zr alloy (aluminium base metal of technical purity).

conditions from base materials of different purity: in one case A999 grade aluminium was used and in the other, aluminium of A8 grade. The samples were obtained at cooling rates of 10^4 K/s, at melt heating temperatures $T = 1270$ K and 1670 K. These conditions correspond to overheating above the liquidus temperature by 100 and 500 K, respectively.

As was noted in Section 2.2.2, under the given solidification conditions primary crystals of zirconium aluminide of the metastable modification form in the alloy; we can change their shape and size by varying the solidification conditions (Fig. 2.16).

The study of morphological features of the forms of growth of zirconium aluminide crystals in the alloy based on commercial aluminium showed that their formation is observed in narrower rate and temperature ranges and that they grow as isotropic faceted crystals reminiscent of a cube in shape. The dendritic growth forms inherent in crystals formed in the Al A999 base alloy are entirely absent.

For clarity, the results described above are presented schematically in Figs. 3.11a,b. Overheating the melt above the temperature T_L ($\Delta T = 250$–300 K) causes an abrupt refinement of aluminides to the size of 1–2 μm in the course of subsequent solidification (the hatched area in Fig. 3.11b). A similar structure in the alloy based on A999 grade aluminium forms at much higher melt heating temperatures $\Delta T = 450$–500 K (the hatched zone in Fig. 2.16c).

To explain the metallographic data, we invoked the experimental results of the study of properties of liquid alloys (Fig. 2.7), which indicate that substituting aluminium of special purity by commercial aluminium leads to a decrease in the temperature of homogenization of the corresponding melt by 350–400 K.

Thus, summarizing the results of the experiments described above, we may draw a conclusion about the generality of mechanisms of the influence on the alloy structure of low-melting additions introduced directly into the melt or present as controlled impurities in the aluminium matrix. These impurities affect the structure of both liquid and solid materials, which gives further evidence of their interrelation. The shift of the homogenization temperature of melts of commercial purity into the region of lower temperatures and the dramatic increase in the number of crystallization centres in this region decrease the chance of creating supercooling at the phase transformation front and bring the system into a more equilibrium condition. According to the crystallization theory [9, 49], the layer-by-layer mechanism of growth of the solid phase is implemented under such conditions and the polyhedral form of growth of crystals with high entropy of fusion becomes the preferential form.

As a consequence, comparison of the results presented in Figs. 2.16 and 3.11 reinforces indirectly the statement made in Chapter 2 that heating of aluminium melts with transition metals or silicon to the region of their homogenization temperatures ($T \leq T_{hom}$) results in the modification of the cast structure of material of any purity.

3.4   Effect of Modifiers on the Structure of Liquid Alloys

In this section we generalize the experimental results concerning the modification of aluminium alloys by various additives and suggest a unified approach to the nature of this phenomenon based on the structural state of their melts.

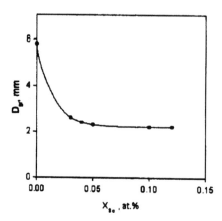

Figure 3.12   Concentration dependences of the mean size of cast grains in aluminium alloyed with scandium.

3.4.1. Cluster mechanism of modification of aluminium alloys

In Section 3.1 we pointed out that in the literature, along with traditional mechanisms of modification of the 1st and 2nd kind, the possibility of implementing alternative ways of using impurities to influence the structure of cast metal is considered. In particular, Samsonov and Lamikhov [18, 37] explained the regular change in the modifying effect with the growth of the number of d electrons in impurity atoms by the nucleation action of impurity clusters of the melt (this kind of heterogeneity was discussed in Section 1.4). According to these researchers, the modifying effect of metals is determined by the reactivity of their atoms, the criterion of which is the degree of filling of the d shell measured by the quantity $1/nn_d$, where n is the principal quantum number of the d shell and n_d is the number of electrons in it. However, the experimental data presented in [37] give evidence that the modifying capability of 3d metals in aluminium rises in the order: Ni → Co → Cr → Mn → Fe → V → Ti → Sc. It is easy to ascertain that the criterion indicated above does not always determine the degree of grain refinement. In particular, the growth of modifying capability from Cr and Mn to Fe does not correlate with decreasing $1/nn_d$.

For a long time, the mechanism of modification suggested in [18, 37] remained a hypothesis. To support this hypothesis, there was little experimental data on the influence of various concentrations of transition metals on the size of primary grain of aluminium and its alloys, as well as inadequate experimental evidence in favour of substantial rearrangements of aluminium melts under the influence of transition metal impurities. Thorough studies of impurity effects in liquid aluminium with additions of these elements created the basis for the study of the influence of cluster inhomogeneity of the melt on the structure of cast metal formed from this melt.

Bazin *et al.* [14] compared the influence of scandium on the properties of liquid aluminium (according to the data [24]) with its modifying action. For this purpose, the grain size (D_{gr}) was examined in specimens containing up to 0.3 at. % Sc. The results are presented in Fig. 3.12. The range of intensive refinement of the structure was found in the $D_{gr}(x_{Sc})$ curve at $x_{Sc} < 0.03$ at. %.

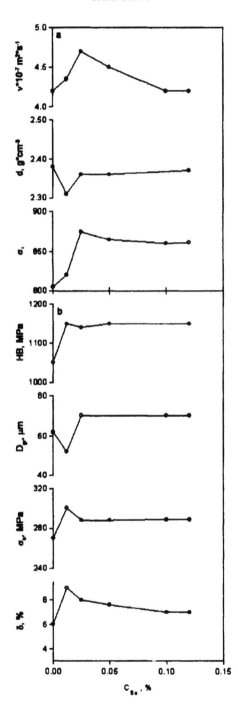

Figure 3.13 Kinematic viscosity ν, surface tension σ, and density *d* at 970 K (a); the mean grain size D_{gr}, ultimate strength $σ_u$, and relative elongation δ (b) for AL9 alloy as a function of scandium content.

Modification of aluminium with scandium at concentrations lower than the eutectic composition ($x_e = 0.3$ at. %) cannot be related to the precipitation of particles of the high-melting intermetallic compound Al_3Sc on solidification, which do not constitute here the primary phase. It can be shown that at low solubility of scandium in solid Al its surface activity at the boundaries between growing crystals and the melt should correlate with activity at the boundary between the melt and vacuum. However, at concentrations below $x_{Sc} = 0.03$ at. %, scandium additive does not influence the surface tension (σ) of liquid aluminium (Fig. 1.13), whereas in the concentration range where σ changes significantly with alloying, the fine-grained structure of the liquid metal remains unchanged. Therefore, scandium cannot be classified as a surface-active modifier. At the same time we noted in Section 1.4 that anomalies in the isotherms of density and viscosity observed after introducing the first portions of scandium point to the formation of clusters containing up to 60 Al atoms. Clearly, in the supercooled melt they serve as nuclei of the solid phase, which results in grain refinement. Thus, the results obtained in [14, 24] gave the first experimental verification of the reality of the modification mechanism proposed in [18, 37].

Owing to the resemblance between the concentration dependences of properties of Al–Ti and Al–V melts and the corresponding curves for the Al–Sc melt (Figs. 1.14 and 1.15), the modifying effect of the first portions of Ti and V was also attributed to cluster formation. However, it remained unclear why transition metals at the end of the 3d series, forming the most extended clusters in aluminium (Table 1.2), are not modifiers of aluminium. The answer to this question was given by Bazin with coworkers [38] who interpreted the results of an X-ray study of Al–Sc, Al–Ti, and Al–Fe melts. They proposed a method for separating the constituents of the integrated scattering curve, which arise due to scattering from impurity clusters and from the matrix, and calculated the characteristics of the short-range order in clusters. It was found that the shortest interatomic spacing and the first coordination number in groups formed by scandium and titanium are close to the corresponding parameters of solid aluminium with an fcc lattice, and that the same characteristics of iron-based clusters, conversely, differ considerably from these. Therefore, only impurities forming in melt clusters with a structure close to the structure of growing crystals possess a modifying capability, because it is such groups that may play the role of nucleation centres of a new phase on supercooling.

More recently, the possibility of modifying aluminium alloys by a cluster mechanism was also shown for multicomponent systems. In particular, Fig. 3.13a shows the dependences of kinematic viscosity (v), surface tension (σ), and density (d) of a commercial AL19 alloy of the Al–Cu–Mn–Ti system on the scandium concentration at the standard temperature of its melting (970 K).

When analyzing these curves, attention should be given to the extrema of v, σ, and d in the region of low scandium concentrations (up to 0.1%). The appearance of the $v(C_{Sc})$ curve is nearly identical to the corresponding dependence for pure aluminium (Fig. 1.14). This led us to conclude that small scandium additives interact basically with Al atoms. It is likely that clusters around Sc ions are also formed in the multicomponent commercial alloy with a scandium concentration of $C_{Sc} \approx 0.04-0.06\%$. When the first portions of the alloying element (up to 0.05%) are introduced, the derivative $d\sigma/dC_{Sc}$ is positive. This means that the impurity is localized predominantly in the melt volume, whereas the surface is depleted of scandium. As the scandium content increases, the surface scandium concentration begins to surpass the bulk concentration ($d\sigma/dC_{Sc} < 0$) and, judging by the viscosity isotherm, clusters form in the volume.

Thus, studies of all the properties of melts of an AL19 alloy with scandium point to significant changes in their structure at $C_{Sc} < 0.1\%$.

The authors compared these results with the influence of scandium on the grain size in ingots of the given alloy obtained by casting into a permanent mould followed by heat treatment (Fig. 3.13b). It was found that introduction of 0.01% Sc causes a substantial decrease in the mean grain size and an enhancement of strength, hardness, and relative elongation of samples.

A comparison of the curves plotted in Figs. 3.13a and 3.13b shows that the most effective grain refinement and strengthening of the AL19 alloy takes place at the compositions at which clusters form in melts; these clusters are likely to be the major reason for the modifying effect of scandium.

The experimental results presented in this section give evidence of the reality of cluster mechanism of modification of aluminium alloys.

3.4.2. Inoculating mechanism

When analyzing the experimental data on the influence of the structure of master alloys with high-melting additives on the formation of the ingot structure, it was noticed that some results cannot be explained by traditional mechanisms of modification of aluminium alloys based on the treatment of aluminium melts as structurally uniform systems. As was indicated in Section 3.1, the modifying effect caused by high-melting additives was most often investigated at concentrations exceeding the peritectic concentrations (0.15% for Ti and 0.11% for Zr). This was thought to be natural since the authors assumed that in the case of heterogeneous solidification of the matrix the high-melting aluminides Al_3Zr and Al_3Ti, initially nucleated in the melt, serve as the crystallization centres.

Our experiments show that the effect of refinement of the as-cast structure is observed also at concentrations lower than the critical one [13, 23].

Analysis of the hereditary relation between the structures of master alloys and the matrix, described in the monograph [25], and the results of the authors [13, 23] show that the inference that the degree of cast grain refinement is inversely proportional to the size of crystals of intermetallics in the master alloy is correct but calls for some refinements and additions. If we compare the data on the efficiency of modification of the structure of ingots of commercial aluminium by zirconium introduced in the melt with the aid of various master alloys – SMA and OMA – then the higher efficiency of the latter cannot be explained because aluminides in both master alloys have the same size and structural characteristics. It is natural to assume that the modification mechanism in this case is related to the structure of melts.

According to the experimental data described in Chapter 2, we may assume that liquid master alloys have a microheterogeneous structure and consist of microregions of various types differing in the content of major components and in the type of short-range order. Along with the basic structural constituent of the melt having composition close to that of

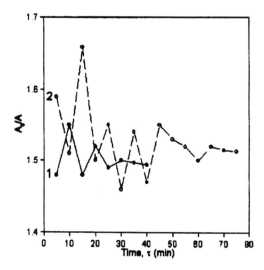

Figure 3.14   Decrement of damping of torsion oscillations derived from viscosity measurements in melts of Al–2% Zr master alloys obtained by different methods: (1) OMA, (2) SMA.

the aluminium matrix, microgroups exist in which the structure of short-range order resembles the structure of aluminides, but the content of the high-melting additive is much higher.

Bearing in mind this assumption, the addition of a high-melting point master alloy to the base aluminium melt should increase its microheterogeneity as a result of the introduction of additional structural and chemical components.

After dissolving solid aluminides in the matrix melt, the microheterogeneous state is retained for a long time at moderate overheating of the system above the liquidus point T_L. Each temperature corresponds to a definite composition, dispersity, and bulk content of colloidal particles which inherited the structure of the intermetallic compound. At subsequent solidification, these particles may play the role of nuclei of the new phase. This explains the strong dependence of the effect of alloy structure modification on the temperature at which the high-melting master alloy was introduced in the initial melt.

Thus, when estimating the modifying capability of master alloys, it is necessary to know the rate of solution of crystals of intermetallic compounds in the matrix melt and to have information on the structural characteristics of the microheterogeneous melt formed after solution.

Adopting the concept set forth above, we may consider the modifying possibilities of master alloys obtained by various methods. Figure 3.14 shows the time dependences of kinematic viscosity of liquid aluminium after introducing, at 1070 K, equal amounts of an Al–Zr master alloy prepared by means of OMA and SMA technologies [13]. These dependencies attest to the substantial relation between the length of relaxation processes and the structure of a master alloy. From these curves we can estimate the incubation period, i.e. the time τ_{crit} elapsed from the introduction of the master alloy in the alloy to the appearance of the modifying effect. The long incubation period for SMA is determined not

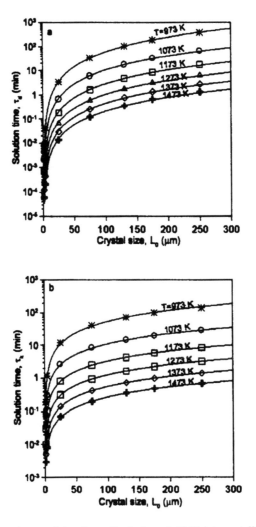

Figure 3.15   Dependence of the time of solution of Al$_3$Ti intermetallics on their initial sizes at different melt temperatures; (a) diffusion regime, (b) kinetic regime.

only by the presence of large aluminide crystals but also by their extremely non-uniform distribution over the ingot cross section. The lesser time corresponding to the introduction of OMA, which points to a higher rate of solution of zirconium aluminides may be caused only by the defect structure of crystals and by the presence in it of finer subgrains of 30–50 μm in size. As long as disperse particles retaining the structure of intermetallic phases exist in the melt, the probability is very high that excess phases of the same composition will appear after solidification. This phenomenon is very often observed in ingot casting and is undesirable because it decreases the placticity of the material. It is clear that such a situation is realized if the holding time after introduction of master alloys is shorter than τ_{crit}. Therefore, in order to assess the modifying ability of master alloys it is important to know the rate of solution of intermetallic phases in the matrix melt.

176   I. G. BRODOVA *et al.*

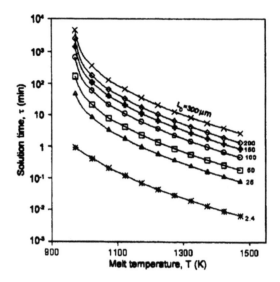

Figure 3.16 Temperature dependence of the time of solution of Al₃Ti intermetallics at different initial sizes of crystals.

This parameter can also be estimated using a calculation method. It is known that the solution of solid particles in a liquid can be represented as consisting of two processes: the transition of atoms from the solid phase into liquid through the surface of interphase boundary and redistribution of the particles in the liquid. Depending on which of these two phenomena determines the rate of the process as a whole, a distinction is made between the kinetic and diffusion regimes of particle solution. From the results of analysis performed by the authors in the work [23], and knowing the size and the bulk content of particles in the master alloy, we can estimate the time of particle solution $\tau = \Sigma \tau_d + \tau_k$, where τ_d and τ_k are the times of solution of aluminides in the diffusion and kinetic regimes, respectively.

Figure 3.15 shows the time of solution of Al₃Ti intermetallic compounds in the aluminium melt in the diffusion (Fig. 3.15a) and kinetic (Fig. 3.15b) regimes as a function of the initial size L_0 of crystals. The temperature dependence of the total time of isothermic solution, τ, for various initial sizes of crystals is presented in Fig. 3.16. We see from the graph that in the diffusion regime titanium aluminides dissolve almost instantaneously ($\tau_d \ll 2$ min). Assuming, on the other hand, that the effects at the interphase boundary play a substantial role in solution, the length τ_k of the process will vary over a wide range, depending on the initial size of crystals. The curves plotted in Fig. 3.16 illustrate clearly the difference between the times of solution of aluminides for crystals of different size and allow us to estimate the rate of solution of crystals of master alloys in the matrix at definite temperatures of alloying the melt. For example, if the master alloy is introduced at 990 K, the solution of 150-μm-sized intermetallics will take 220 min, whereas the solution of 3-μm-sized crystals, only 1.2 min.

Although such estimates were made for crystals of titanium aluminide, we may assume that the character of T-dependence of τ is also retained for zirconium aluminides, since these aluminides have similar physical properties.

Knowing the temperature dependence of the time of dissolution of aluminides, we can explain correctly the behaviour of one or another master alloy under various alloying conditions. It is clear that Al–Ti and Al–Zr HMAs, having small-sized intermetallics, possess good modifying properties even at low temperatures, while the modifying capability of SMAs is realized successfully only at elevated temperatures, and complete solution (due to the large size of aluminide crystals) requires either high overheating above the liquidus or rather long isothermal holding near the liquidus, which cannot always be achieved in practice.

Moreover, when estimating the modifying effect of HMA, we should take into account an additional positive contribution of the crystallographic factor. Since disperse particles inheriting the crystal structure of intermetallic phases persist in the melt after solution of solid aluminides, in the case when we use a master alloy of this type these particles have a short-range order structure close to the cubic symmetry of the aluminium matrix. In this situation, the efficiency of structural constituents of the melt as crystallization centres is determined by the principle of structural and dimensional conformity developed by Dankov, which suggests that the activity of crystallization centres depends on the conjugation and proximity of lattice parameters of particles serving as nuclei and of the lattice parameters of the solid phase formed on these particles [6]. This fact is easily established if we turn to the data presented in Table 3.8. The values of the structural misfit coefficients (K_{str}) in this table illustrate clearly the higher efficiency of aluminium matrix modification by master alloys containing intermetallic compounds of metastable modification. Thus, the above facts give evidence of the inoculation mechanism of modification of the structure of aluminium alloys with zirconium and titanium, which consists in creating additional crystallization centres owing to the increase in microheterogeneity of the melt after introducing high-melting master alloys.

Table 3.8 Structural characteristics of intermetallic compounds.

Stoichiometric composition of intermetallics	Lattice type	Lattice parameter a, nm	$K_{str}*$
Al	cubic	0.40496	0
Al$_3$Ti	tetragonal, DO_{22}	0.3835, c/a = 2.23	5.3
Al$_3$Ti	cubic, $L1_2$	0.4041	0.24
Al$_3$Zr	tetragonal, DO_{23}	0.4306, c/a = 3.92	6.3
Al$_3$Zr	cubic, $L1_2$	0.4050	0.01

$$*K_{str} = \frac{a_{phase} - a_{Al}}{a_{Al}} \times 100\%.$$

3.4.3. Adsorption mechanism

Joint consideration of experimental data concerning the influence of low-melting additives on the properties of liquid alloys and on the structure formation in the course of solidifi-

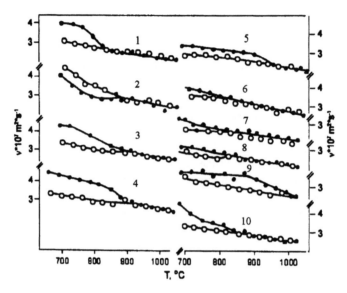

Figure 3.17   Temperature dependences of viscosity of Al–5.4 at. % Sn alloy (1) with Mg (2), Cd (3), Ti (4), Mn (5), Sc (6), Gd (7), Zr (8), B (9), and Zn (10) additives; (• heating, ○ cooling).

cation of these alloys has shown that their modifying effect is determined by the processes of structural rearrangements in the melt [20, 35, 39, 42].

According to the conclusions of the theory of metastable microheterogeneity, the thermal stability of the microheterogeneous state is determined chiefly by the interphase tension σ at the boundaries of fine-grained particles (see Section 1.8). As a consequence, the introduction into the melt of minor additions of substances lowering the σ value (so-called surface-active substances) can decrease markedly the temperature of its homogenization. This makes it possible, in turn, to perform heat treatment of liquid metal under much less severe conditions or, in some cases, makes it generally redundant.

Above we noted the facts of decreasing temperature of transition of melts into the state of a true solution with increasing impurity content (see, for example, the data on viscosity of Al–Zr melts in Section 2.1).

While in [20] the set of impurities was chosen by accident, later on, in the work [39], we turned to the choice of surface-active substances using the known criteria of surface activity [43, 46].

We chose Al–5.4 at. % Sn as a base alloy, whose temperature dependences are shown in Fig. 2.5a. Attention should be paid to the distinct discrepancy between the heating and cooling curves which have a branching point near 1270 K, indicating that the transition of the melt from the metastable microheterogeneous state into the true solution state occurs near this temperature. The phase constituents of the metastable colloid in this system are diluted liquid solutions of aluminium in tin and tin in aluminium, which in the first approximation can be identified with pure components.

According to Antonov's rule, in the case of good wettability the interphase tension σ_{Al-Sn} at the boundary between the colloidal particle and the dispersive medium can be estimated as the difference between the surface tensions of these phases at the vacuum boundary: $\sigma_{Al-Sn} = \sigma_{Al} - \sigma_{Sn}$. The introduced additives either do not influence the σ_{Al} and σ_{Sn} values (inactive impurities) or lower the surface tension (surface-active substances). We may assume therefore that substances which are surface-active in liquid aluminium and inactive in liquid tin are the candidates for the role of surface-active substances at the interphase "tin–aluminium" boundary, i.e. impurities reducing thermal stability of the colloidal state in Al–Sn melts.

The surface activity Q of the dissolved substance at the interface considered here is determined as $Q = (\partial\sigma/\partial c)_{c\rightarrow 0}$, where c is the concentration of the given component. Several criteria of surface activity, systemized in the monograph [46], were suggested. Their analysis shows that in spite of the fact that various criteria do not always agree with each other such elements as Mg, Cd, Zn, Gd, Sc, Zr, Ti, Mn, and B are possible candidates for the role of surface-active substances in Al–Sn melts existing in the metastable microheterogeneous state.

In order to compare the efficiency of the influence of various additives on the homogenization temperature T_{hom}, identical amounts (0.14 at. %) of these additives were introduced into the matrix melt. The results are presented in Fig. 3.17. In all cases, except for the samples containing Cd and Sc, the $v(T)$ curves obtained in heating and subsequent cooling have clear-cut branching. The introduction of the above indicated amounts of Mg, Cd, Zn, Zr, Mn, or Ti additives changes noticeably the appearance of the temperature dependence of viscosity but influence only slightly T_{hom}. In contrast, additives of B and Cr markedly raise the homogenization temperature. It was found therefore that the criteria of surface activity do not determine by themselves the effect of additives on interphase tension and, consequently, on thermal stability of the metastable microheterogeneous state.

One of the reasons for this may be the non-monotonic concentration dependence of the interphase tension σ at the boundary of disperse particles similar to those discussed in Section 1.3. Evidently, for significant reduction of σ, and therefore the homogenization temperature, it is important not only to choose correctly the surface-active additive but also to determine its optimal concentration. This reasoning is strengthened by the results of experiments in which the effect of scandium content on T_{hom} of the Al–5.4 at. % Sn was examined.

The results presented in Fig. 3.18 point to the substantial dependence of the homogenization temperature on the concentration of surface-active substances. Indeed, the introduction of only 0.05 at. % Sc in the initial melt decreases T_{hom} down to 1000 K, i.e. to the standard temperature of melting aluminium alloys in industrial conditions. However, an increase in x_{Sc} raises the homogenization temperature to nearly 1250 K. With further increase in the scandium content, the values of T_{hom} again decrease reaching 1060 K at $x_{Sc} = 0.5$ at. %.

Thus, the criteria of surface activity give only a rough idea of additives that must be introduced into microheterogeneous melts to lower the interphase tension at the boundaries of colloidal particles and the temperature of transition of the system into the true solution state. This preliminary choice should be followed by optimization of the concentration of surface-active substances in the melt.

The only class of liquid metallic alloys for which the interphase activity of the impurity can be determined experimentally are systems with a miscibility gap. The existing

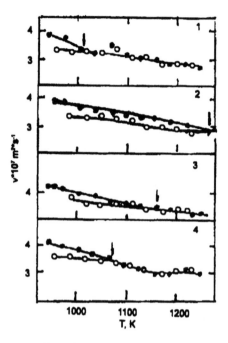

Figure 3.18 Temperature dependences of viscosity of Al–5.4 at. % Sn alloy with scandium addition; (1) 0.05 at. % Sc; (2) 0.14 at. %; (3) 0.30 at. %; (4) 0.50 at. %. Arrows mark T_{hom}; • heating, ○ cooling.

procedures for measuring interphase tension enable one, by performing experiments in the temperature range below the segregation point, to explore the effect of the impurity on σ and then to use the results obtained for predicting the efficiency of introducing this impurity with the aim of decreasing the temperature of homogenization of the microheterogeneous melt formed after going beyond the bounds of the immiscibility region.

It is precisely this approach that was used by Chikowa *et al.* [42]. Using the viscosimetric method for measuring homogenization temperatures of Al–Pb melts with various amounts of tin, the authors of [42] established that the state of true solution can be attained without significant overheating of the melt above the immiscibility dome even at 0.5 at. % Sn. In the absence of a surface-active impurity this state was obtained only at temperatures of 200–250 K above the dome.

The results presented in this section show that in some cases the optimal choice of the surface-active substance and its concentration provides a significant decrease in the homogenization temperature of microheterogeneous melts.

In the experiments described above, the reduction of T_{hom} of microheterogeneous melts caused by microalloying with some additives was determined directly by the shift of the branching points the temperature curves of viscosity toward the region of lower temperatures. This effect is indirectly evidenced by the data on modification of rapidly cooled silumins by low-melting Zn and Sn additives (see Section 3.3).

The results presented in Chapter 2 show that knowing the structure of rapidly quenched materials we can supposedly judge the processes occurring in the melt on heating. For example, the dependence of the volume fraction of crystals, their size and morphology on the temperature of heating the initial melt points indirectly to the degree of overheating and the number of solidification nuclei in the melt. These parameters are controlled by the complex processes of solution and dispergation of structural inhomogeneities, which occur in melts upon heating. Let us consider in this context the results of quantitative metallography of rapidly quenched specimens of a hypereutectic Al–26% Si silumin before and after alloying it with tin [35] (Figs. 3.6a,b). The transformation of temperature dependences of the volume fraction of crystals, which was revealed in alloys with Sn additives, shows that the introduction of such additions changes the structural state of the initial melt. In connection with this, the increase at $T = 1370$ K in the volume fraction of silicon crystals, observed for the alloy with Sn additives, and its subsequent fall with increasing melt heating temperature, is of particular interest.

The same experimental fact was found in the solidification of Al–Zr and Al–Ti melts rapidly cooled from temperatures close to the temperatures of their transition from the microheterogeneous to homogeneous state [50]. It may be assumed that in an Al–26% Si melt alloyed with tin or zinc the homogenization temperature lies near 1370 K. It was found that for the binary Al–26% Si composition this temperature is much higher (1650 K [45]). Thus, the introduction of Sn microadditions into the silumin melt resulted in the decrease of T_{hom}.

Let us consider this phenomenon within the framework of the colloidal model of the structure of eutectic melts described in Chapter 1. It may be thought that on account of the small quantities (< 1 at. %) and weak bonds (low melting temperatures), these microadditions, upon introduction into the silumin melt, do not participate independently in creating colloidal particles that may become the centres of subsequent solidification. This suggests that the effect of microadditions in hypereutectic silumins is concentrated on already existing nuclei – Si-rich particles. Based on the experimental data [45], showing that hypereutectic Al–Si melts represent a colloidal suspension of disperse silicon particles in an Al-rich melt, we related the mechanism of action of the additive to its surface activity with respect to silicon, assuming in this case the existence of parallelism in the change of surface tension at the melt–vapour and melt–solid body boundaries [46].

As a surface-active additive lowering the surface energy of a Si-rich colloidal particle, tin reduces the thermal stability of the colloidal melt. On adopting such a mechanism of influence of surface-active additives on liquid silumins, we can explain correctly the relation between the modifying effect and heat treatment of the melt. The basis for this reasoning are the experimental results described in Chapter 1, which indicate that the dimensions of colloidal particles fall with increasing a melt heating temperature, while at temperatures close to T_{hom} their slow solution in the kinetic regime may alternate with comparatively rapid processes of consecutive dispergation of colloidal particles. It may be assumed that in aluminium melts containing high-melting components (for example, Si), on heating above a definite temperature a transition occurs from the microheterogeneous state to a different, but also microheterogeneous, state with a more disperse structure. Such a rearrangement initiates in turn a spontaneous rise of crystallization centres, which manifests itself in modification of the silumin structure.

Thus, the introduction into melts of hypereutectoid silumins of low-melting additives, surface-active at the boundaries between silicon and liquid aluminium, governs the amount and sizes of the structural constituents and may exert, in some HTT regimes, a strong modifying influence on the structure of material in the solid state.

To summarize, we should note that the entire body of experimental data indicates that both low-melting and high-melting additives affect the state of aluminium melts by changing the distribution, amount, and size of structural constituents of the melts. At definite temperatures and concentrations of these additives, a drastic dispergation of colloidal particles occurs in melts, which results in an increase of the number of nucleation centres during the course of subsequent solidification. In the case of alloying with surface-active additives, the cause of this phenomenon is the increase in number and decrease in size of disperse particles in the microheterogeneous melt. If we use, however, high-melting master alloys, the above phenomenon is related to the introduction of additional chemical inhomogeneities inheriting the structure of aluminides.

From this point of view, all the additives considered above have a modifying effect of an inoculation type on the structure of alloys. In such an approach, there is no need for the classical division of all existing modifiers into I and II type modifiers and a unified mechanism of modification of the as-cast structure of alloys is laid through the action on their melts.

Moreover, if we add to this reasoning the impurity mechanism of modification (Section 3.4.1), which is also related to the action of microadditions on the melt structure, it becomes evident that the effect of structure refinement in aluminium alloys is due largely to the modification of their structure in the liquid state.

3.5   Microalloying as a Tool for Improving Properties of Some Commercial Alloys

3.5.1. Deformable alloys

Deformable aluminium-based alloys constitute a large group of structural materials which are of prime importance in various fields of industry. A great many works are devoted to improvement of the structure and upgrading service characteristics of materials; the most important are the works in which the interrelation between the as-cast structure and the structure of semi-finished products is shown [17, 26–29, 31, 48].

While in recent years the number of experimental studies on the influence of HTT of melts on the structure and properties of melts has been continuously growing, the way of improving the structure by means of various modifying additives remains traditional in industrial practice, since they provide the most extensive spectrum of new properties.

As was indicated above, progress in the study of metallic melts, as well as regularities revealed in solidification of alloys with many additions, pointed to new directions in this seemingly thoroughly investigated area. One of the ways is to search for new technologies for producing multicomponent alloys providing high service characteristics owing to the combined use of microalloying with high-melting additives and heat time treatment of their melts.

Approaches to the development of scientifically justified technologies with the use of HTT of melt and modification, which were worked out on model systems, were tested on a series of commercial deformable alloys relating to various systems: Al–Cu (2324),

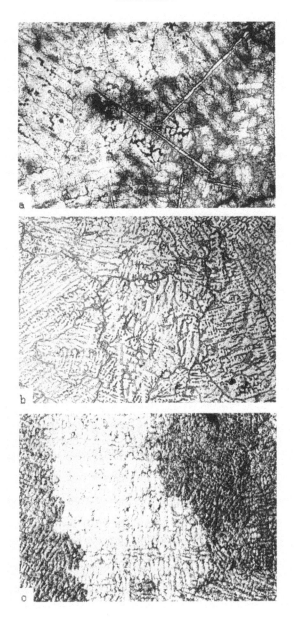

Figure 3.19   Macrostructure of AK6 ingot alloyed with 0.1% Ti with the use of different master alloys: (a) SMA, (b) HMA, (c) OMA (×80).

Al–Mg (5050), Al–Li–Mg (5XXX), and Al–Cu–Mg–Si (2017)*.

Without going into detail of the experiment described in [23], we present the results on alloying the AK6 (2017) alloy (with 2% Cu, 0.6% Mg, 0.6% Mn, 1% Si, the rest being Al)

---

* The alloy grades by the ASTM standard [48] are indicated in parentheses.

used for stampings and forgings. Alloying with high-melting additives was used in this case with the aim to raise the strength characteristics of materials with conservation of high plasticity under hot conditions. The implementation of these properties required a uniform disperse cast structure of the α-solid solution with maximum content of alloying additives. This required, in turn, a master alloy possessing high modifying and alloying capabilities.

For this purpose, master alloys with identical composition (Al–3.8% Ti), obtained with the use of HTT of melt at low and high cooling rates (OMA and HMA, see Section 3.2) were employed. For comparison, an AK6 alloy with titanium additive, introduced with the aid of pig master alloy (SMA), was also investigated. Analysis of the results showed that the structure of the AK6 alloy depends strongly on the type of the master alloy used (Figs. 3.19a–c, Table 3.9).

Table 3.9   Grain size (D_{gr}) and microhardness (H_μ) of AK6 alloy with 0.15% Ti addition introduced with the aid of master alloys of various types.

Master alloy type	D_{gr}, μm	Presence of intermetallics	H_μ, MPa	Modification factor, K_m*
SMA	200	yes	900	1.3
HMA	20	no	1040	13
OMA	120	no	1120	2.2
–	260	–	870	1

*Here, and henceforth we use for the modification factor (K_m) the ratio of the mean linear grain sizes in alloys before and after alloying with various additives.

For example, when SMA is used in the casting process, a strong difference in grain sizes is observed in the ingot structure and excess primary Al_3Ti intermetallics are present in the form of plates of 200 μm length and 10 μm width crystallizing in a tetragonal DO_{23}-type lattice (Fig. 3.19a).

In the specimens obtained with the aid of HMA, there is no primary excess phase, grains have a subdendritic morphology, their size is reduced by a factor of 10 (Fig. 3.19b), and the H_μ values are increased by 80 MPa (Table 3.9). If the OMA alloy is used as a modifier, the crystallization of the solid phase has a dendritic character and grains of the α-phase are smaller by a factor of two than grains in samples alloyed with titanium with the aid of SMA (Fig. 3.19c). High H_μ values and the absence of primary intermetallics also give evidence of the solidification of the α-solid solution, which is strongly oversaturated with titanium.

As we have often pointed out, one of the most important factors influencing the efficiency of master alloys is the choice of conditions for their introduction into the base melt. In this experiment, the master alloys were introduced into the liquid alloy at $T = 990$ K, whereupon the melt alloyed with titanium was soaked for 15 min. and poured out into alundum crucibles. The assessment of the temperature-time regime of preparation of the melt for pouring, based on the calculation of kinetics of solution of intermetallic phases of various sizes (see Section 3.4), as well as the available experimental data on the properties of alloyed melts, proves that the cause of formation of different structures lies in the different structural states of liquid metal. In particular, the appearance of excess crystals in ingots alloyed with SMA is due to the retention in the melt of undissolved aluminides, whereas the formation of a great number of disperse subdendritic grains in an alloy with titanium additive introduced with the aid of HMA is determined by the special state of the melt, which at the given HTT regime was a microheterogeneous system containing a great number of potential effective crystallization centres. The lack of excess phases and the high degree of alloying of the α-solid solution in the structure of ingots produced with OMA indicate that defects in the aluminide structure facilitate rapid dispergation and solution of aluminides in the matrix melt.

Table 3.10 contains the values of mechanical characteristics of alloys with titanium additives introduced with the aid of various master alloys after standard heat treatment (quenching + artificial ageing).

Table 3.10   Mechanical characteristics of specimens made of the AK6 alloy with titanium additive.

Master alloy type	σ_u, MPa	δ, %
–	360	5
SMA	380	5
OMA	400	7
HMA	420	10

Thus, we can show clearly that the character of the macrostructure and the level of properties of an alloy can be varied over a wide range by the proper choice of a master alloy.

Similar experiments were also conducted with some deformable alloys with zirconium additives. Their goal was to refine cast grains and to increase the degree of alloying of the α-solid solution, i.e. to combine the functions of alloying and modification of alloys in a single technological cycle. A comparison of the efficiencies of two Al–2% Zr master alloys obtained with the use of SMA and HMA technologies showed unambiguously the advantage of the latter alloy.

The increased requirements for the production of ingots with ultrafine homogeneous structure stimulated a search for new effective modifiers of aluminium alloys, one of which is scandium [26, 44]. We investigated the interrelation between the structures of liquid and solid alloys doped with scandium with the aid of a binary Al–Sc master alloy. As a result, the modification conditions were optimized and the structure of the material was improved. We used the AMG6 alloy as an object for investigation.

The regimes of alloying with scandium were chosen with allowance for the structure of the initial melt. For this purpose, we obtained the temperature dependences for viscosity of the AMG6 alloy in the liquid state before and after alloying with scandium in amounts of 0–0.4%. Without going into details of the experimental method, we can indicate the following results obtained in the study of properties of liquid alloys (Figs. 3.20a–e). Disagreement was found between the temperature dependences of heating and cooling, i.e. we observed hysteresis that can be related to the destruction of the microheterogeneous state of the melt. The metastable heterogeneity of the melt is due mainly to the hereditary effect of the structure of charge materials. Perfect mixing of the components on the atomic level is observed, according to the data of this experiment, upon heating the melt above T_{hom} = 1020 K. The introduction of scandium into the melt in amounts of 0.1–0.4%

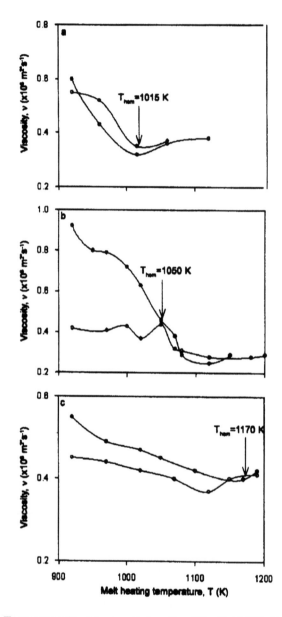

Figure 3.20   Temperature dependences of viscosity of liquid AMG6 alloy with different
scandium additions: (a) 0%; (b) 0.1%; (c) 0.3%.

enhances the melt heterogeneity and shifts T_{hom} toward higher temperatures, for example,
T_{hom} reaches 1185 K for the alloy with 0.4% Sc.

The temperature–time regimes of alloying were chosen on the basis of the experimentally determined viscosity anomalies of melts. The structure of the initial ingot is presented
in Fig. 3.21a. The grains have a weakly pronounced dendritic structure: only the dendritic
contours of grains are retained, whereas primary columns and higher-order branches are not

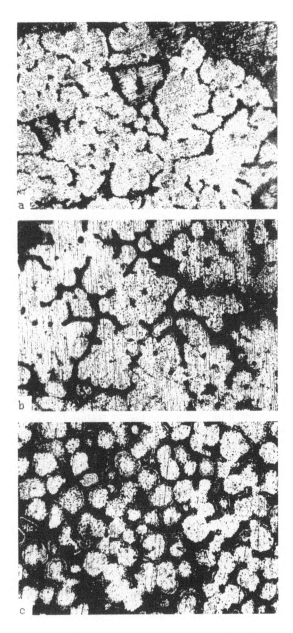

Figure 3.21   Structure of AMG6 alloy ingot before (a) and after alloying with 0.2% Sc (b) and with 0.3% Sc (c) at $T \leq T_{hom}$ ($\times$ 80).

resolved explicitly. The variation in the ingot grain size as a function of scandium concentration and HTT regimes is presented in Fig. 3.22 and in Table 3.11; their macro-structure is shown in Figs. 3.21b,c.

As follows from metallographic data, grain refinement is accompanied by degeneracy of dendritic forms of growth of the solid phase, and by a decrease in the degree of dendritic

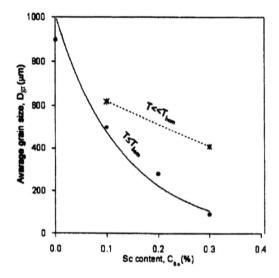

Figure 3.22   Dependence of the mean cast grain size in AMG6 alloy on scandium concentration and HTT of melt.

segregation which shows up in dispergation and reduction of the volume fraction of phases of segregation origin. The results presented in Table 3.11 and in Fig. 3.21 reveal unambiguously the important role of HTT of the melt in forming the cast structure of the alloy. A more disperse structure was found in the ingots that were melted after preliminary high-temperature treatment of the melt following the regime $T \leq T_{hom}$.

Table 3.11   Chemical composition, casting regimes, and grain size in ingots made of the AMG6 alloy with scandium addition.

Sc content, %	Regime of HTT of melt	Modification factor, K_m
	$T \leq T_{hom}$	1
0.1	$T < T_{hom}$	1.3
0.1	$T = T_{hom}$	1.8
0.2	$T \leq T_{hom}$	3.5
0.3	$T < T_{hom}$	9
0.3	$T \ll T_{hom}$	2.5
0.3	$T \leq T_{hom}$	11

With allowance for the general structure of aluminium melts with 3d metals, we can explain these results on the basis of the hypothesis previously proposed by us on the increase near T_{hom} of the number of potential crystallization centres – complexes with a short-range order inheriting the structure of the initial Al_3Sc aluminides.

Thus, we used high-temperature treatment of the melt in combination with alloying with scandium in amounts of 0.2–0.3% as an effective tool for improving the ingot structure (Fig. 3.22).

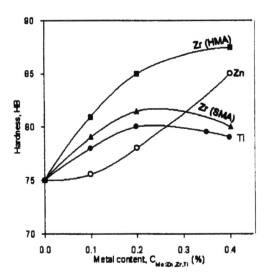

Figure 3.23   Dependence of hardness of castings of AK7M3 alloy on concentration of alloying additives.

3.5.2. Casting alloys

The development of powerful modern industries calls for the design of new materials possessing improved service characteristics, such as strength, hardness, and heat resistance. To solve these problems, technologies of producing castings have been improved and new approaches to alloying and modifying the structure of materials are being developed in materials science [47].

It is worth noting that modification as a technological procedure is most often used for controlling the properties of deformable alloys, whereas it is less well understood and more rarely used for casting alloys. In the recent years, in connection with the increased production of secondary alloys and extension of their industrial application, this line of investigation has assumed particular importance.

We conducted a cycle of studies of the effect of microalloying with various additives on the structure and properties of AK7M3 (A380) silumin alloyed with Mg and Cu. Both low-melting (zinc) and high-melting (zirconium, titanium) elements were used as additional alloying components. The latter were introduced with the aid of master alloys of SMA and HMA type, the preparation and structure of which are described in Section 3.2. To determine the optimal parameters of alloy cast technology, we varied the temperature of introducing master alloys in the melt and the time of holding the melt before solidification. With the aim of optimizing the amount of additives introduced into the alloy, their concentration varied from 0.1 to 0.4%.

Figure 3.23 shows the hardness of castings (sand casting) as a function of the additive concentration in the alloy. We may assume that different character of this dependence suggests a different mechanism of the influence of additive on the strengthening of material. In particular, zinc goes over completely into the α-solid solution, which is evidenced by the reduction of the lattice parameter by 0.3 nm and the absence of additional

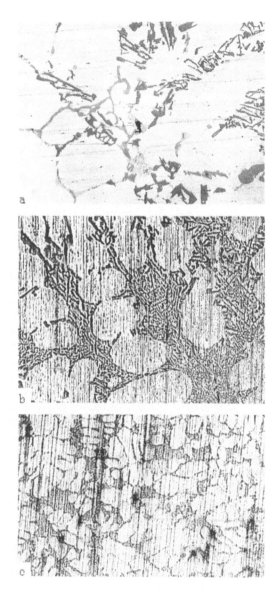

Figure 3.24 Structure of sand (a, b) and chill (c) castings of AK7M3 alloy after alloying with zirconium with the use of SMA (a) and HMA (b, c) (× 200).

intermetallic phases on diffraction patterns. Therefore, the mechanism of the influence of zinc on hardness is a solid-solution hardening.

The dependence of hardness on titanium concentration is linear up to $C_{Ti} = 0.2\%$. The absolute hardness growth is 5 HB. Electron microprobe analysis of the structure of castings with $C_{Ti} > 0.2\%$ showed the presence of primary Al_3Ti intermetallic compounds located at random inside the dendrites of the α-phase. In this case, assuming the optimal regime of melt preparation, the formation of such crystals is determined only by the phase diagram of the alloy. These crystals are an undesirable element in the structure of castings in view of their high brittleness.

Interesting results were obtained with zirconium as alloying additive. We call attention to the different character of dependences of the Brinell hardness HB on zirconium concentration in the case when zirconium was added in the form of master alloys obtained with the use of different technologies. This serves as yet more proof of the existence of a hereditary relation between the charge structure and the structures of the melt and the casting obtained from the latter. When SMA is used as a master alloy, the character of this dependence supports the regularities obtained in the case of alloying with titanium: th e linear growth of hardness by 6–7 HB is observed up to zirconium concentrations > 0.2%. In this case, metallographic analysis points to the presence of excess crystals of zirconium aluminides Al_3Zr with a tetragonal lattice of the $D0_{23}$ structural type (Fig. 3.24a).

In another series of experiments, where Zr was added into the alloy with the aid of HMA, the following results were obtained. No excess intermetallic Al_3Zr phases are revealed in the structure of castings, and the eutectic consists of the disperse silicon phase distributed uniformly between dendrites of the α-solid solution (Fig. 3.24b). As follows from the data presented in Fig. 3.23, the increase in zirconium content in the alloy is accompanied by an increase in its hardness. The maximum increase in hardness of the material by 13 HB was obtained in castings containing 0.4% Zr. Moreover, if we compare the hardness of two materials with the same composition, alloyed with the use of different Al–Zr master alloys, we will make sure that the hardness of the material obtained with HMA is higher by 4–5 HB. We can indicate two reasons leading to this result: solid-solution hardening of the α-phase, which proves to be more strongly alloyed due to the absence of excess aluminides, and the presence of a modified cast structure with a mean grain size decreased by a factor of three.

To obtain a more accurate idea of the modification mechanism, we carried out quantitative EPMA of the distribution of alloying additions over the grain size in the central part of grains. As a result, local 1–2 μm regions were found with increased Zr and Ti content. According to the EPMA data, they have the composition of $Al_5(TiZr)$ aluminides. The mechanism of formation of these phases needs clarification but we may assume that they result from the primary crystallization of intermetallic phases inheriting the structure of the multicomponent complexes making up the microheterogeneous melt. The presence of these phases ensures the uniform disperse as-cast structure of castings and high hardness of the material.

The combined action of Zn and Zr additives, as well as Zn, Zr, and Ti on the structure and hardness of the AK7M3 alloy with the use of the modifying compositions 0.2% Zr + 0.4% Zn and 0.2% Zr + 0.4% Zn + 0.2% Ti did not significantly change the picture. The overall increase in hardness as compared with the initial alloy was 10 HB. This fact points to the nonadditivity of the three elements in their action on hardness of the material and to the complex mechanism of its alloying.

The effect of modification of the structure of castings by microalloying was tested on specimens cast in a permanent mould. The structure of such castings is shown in Fig. 3.24c. Comparison of the structures in Figs. 3.24b,c shows their susceptibility to solidification conditions. Modification results in threefold refinement of the dendritic cell of the α-phase and in the substantial refinement of crystals of eutectic silicon and intermetallic phases of segregation origin. The hardness of such a casting before alloying was 85–88 HB; after alloying with zirconium in amounts of 0.1 and 0.2% it increased to 92–94 HB and 98–100 HB, respectively. The cast grain size decreased by a factor of 4.5–5 as a result of Zr alloying.

From the results of these investigations we established the optimal compositions of secondary silumins and regimes of casting some components for internal-combustion engines.

Analysis of the experimental material presented in this section leads us to a conclusion about the obvious advantage of aluminium-based master alloys with high-melting additives (Zr, Ti, Sc) worked out by the authors, as compared to master alloys cast into pigs.

Moreover, we demonstrated additional possibility of modifying the structure and improving properties of casting, deformed, and granulated alloys by means of HTT of their melts in the process of alloying with high-melting additives.

Chapter 4

Ultrasonic Treatment of Aluminium Melts as an Alternative to HTT for Improving the Quality of Castings and Deformed Semi-finished Products

In sections 1.5 and 1.6 we showed the principal possibility of improving the structure and properties of cast and deformed metal as a result of various external actions on the micro-heterogeneous melt. Of special significance among such actions is the treatment of a liquid metal by powerful ultrasound accompanied by the development of acoustic cavitation and acoustic flows in the metal volume. In some cases the effect of cavitation treatment of liquid aluminium alloys is much more advantageous (more rapid and more effective) than the similar effect produced by heat–time treatment (HTT) of melt.

It seems that Frenkel was the first who called attention to these facts [1]. Analyzing the thermodynamic conditions for the formation of short-range order in liquids as a function of temperature, Frenkel indicated that most researchers ignore the effect of the external pressure on liquid. Frenkel writes: "The possibilities of such an external action typically meet the difficulty of experimentally obtaining negative pressures required for all-round extension of the liquid". Cavitational treatment of melts, i.e. creation in them of cavitation bubbles arising and fragmenting into smaller bubbles at a rate equal to the ultrasound frequency by a chain reaction mechanism, *produces unique possibilities facilitating the effective homogenization of melts* [2–4].

In this Chapter we consider the theory and practice of cavitation treatment of melts with the aim of controlling the structure and improving the properties of cast and deformed metal.

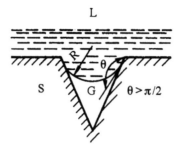

Figure 4.1   Frenkel–Harvey model of a cavitation nucleus in the form of an unwettable (hydrophobic) solid particle in liquid.

4.1   Specific Features of Initiation and Development of Acoustic Cavitation and Acoustic Microflows in Aluminium-based Alloys

Among the physical effects accompanying the propagation of high-power ultrasonic oscillations in a fluid (a melt), cavitation, i.e. the formation of the ensemble of bubbles, almost empty or filled with gases dissolved in the melt is most important. These bubbles are formed by tensile stresses created by the sound wave at the half-period of rarefaction (negative pressure) and continue to grow by inertia for some time. However, the bubbles collapse during the subsequent compression (positive pressure) half-period, thus producing high-intensity shock waves and microflows in the fluid.

Consequently, in an ultrasonic field, gas bubbles are produced at the weakest points of the melt as a result of short-time pressure reduction, and collapse during intervals of increased pressure. These processes occur constantly in the course of high-intensity ultrasonic treatment with the frequency of the ultrasound applied (normally with a frequency of 18 kHz).

It is this unique method that controls the liquid disruptions with a rate of applied sign-alternating pressure, thereby providing the best implementation of Frenkel's idea of intensification of the formation of short-range order in liquid.

Acoustic cavitation and the energy expended on its formation and development give rise to additional heating of the melt. They are responsible for the formation of acoustic flows in liquid metal treated by ultrasound. The latter factor should also favour melt homogenization.

4.1.1. Cavitation nuclei and non-metallic inclusions in melt

Theory and experiment show that some threshold value of sound pressure must be applied to the melt to initiate cavitation or, as is customary to say, to overcome the cavitation threshold. It is known that liquids of very high purity suffer disruptions at rather high negative pressures of the order of 1000 MPa. In particular, for pure water (the most widely studied liquid, from an acoustical point of view), the fracture strength is close to 1000 MPa. However, according to the generalized data obtained by Sirotyuk [5], in normal tap water the cavitation threshold is reduced to 0.1–1.0 MPa.

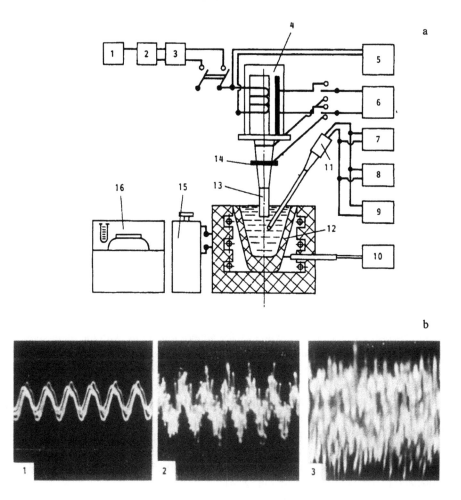

Figure 4.2 (a) Schematic representation of an arrangement for studying the conditions for acoustic cavitation in liquid aluminum and its alloys. (1–3) ultrasonic generators, (4) magnetostrictive transducer, (5) frequency meter, (6) feedback voltage meter, (7) oscilloscope, (8) chart recorder, (9) cavitometer, (10) potentiometer with thermocouple, (11) waveguide sensor, (12) crucible filled with melt, (13) ultrasonic radiator, (14) amplitude sensor, (15) voltage regulator of the melting furnace, (16) hydrogen analyzer. (b) Typical spectrograms of the main frequency signal (18 kHz) before cavitation (1), at the cavitation threshold (2), and in developed cavitation (3).

As was theoretically predicted by Frenkel [1] and proven experimentally by Harvey, McElory and Whiteley [6], *a decrease in the cavitation fracture strength is determined by the presence in liquid of non-wettable (hydrophobic) solid particles which have surface fractures unfilled with gas dissolved in liquid* (Fig. 4.1). Frenkel emphasizes that the degree of adhesion or wettability at the surfaces of particles present in liquid, rather than the size of these particles, determines the capability of particles to become the nuclei for cavitation phenomena.

Calculations and experimental estimates of the cavitation strength for liquid metals show that the cavitation threshold in metallic melts is higher than in water by an order of magnitude. This is related to the difference in physicochemical characteristics and to the specific features of binding of gaseous and non-metallic inclusions in a liquid phase.

Real liquid metal is by no means a perfect liquid: it contains a large amount of insoluble impurities. So, its cavitation strength in the liquid state, just as the structure of the ingot (casting) upon solidification, is determined to a great extent by the purity of the metal with respect to non-metallic impurities. An additional factor affecting the cavitation threshold is the gas dissolved in the melt, which may go over from the dissolved state into cavitation bubbles and back at various phases of pressure oscillation. As a consequence, we should analyze the conditions favouring the nucleation of cavitation bubbles in the complex system consisting of the melt, non-metallic inclusions, and gas.

In our case we will consider the system including liquid aluminium, aluminium oxide, and hydrogen, because it is these non-metallic impurities that determine primarily the purity of liquid aluminium alloys. The problem of determining the cavitation threshold in this system is also complicated by the high chemical activity of the melt.

The method used by us in [2–4] for determining the cavitation threshold in liquid aluminium and its alloys is based on the qualitative and quantitative control of cavitation noise. It is known that cavitation in liquid, i.e. the formation and collapse of cavitation bubbles of various sizes, is accompanied by the generation, along with the main signal of carrier frequency, of acoustic harmonics, subharmonics and incoherent noise. The latter owes its origin to the processes of bubble collapse and the formation of shock pulses and microflows. Figure 4.2a presents the block diagram of the experimental setup for studying the conditions for cavitation in the aluminium melt. Ultrasonic oscillations are introduced into the volume under investigation. The waveguide noise caused by cavitation is displayed and fixed at the oscilloscope screen (Fig. 4.2b). The time of transition across the cavitation threshold is fixed by its appearance. The degree of this process is estimated with special devices – cavitometers. A detailed description of the characteristics of this measuring technique and procedure can be found in [2–4].

The significant influence of non-metallic inclusions in the melt on the formation of cavitation is evidenced by the curves shown in Fig. 4.3: when the content of hydrogen and Al_2O_3 varies in commercial aluminium, the cavitation threshold responds only slightly to the fourfold rise in the hydrogen concentration. At the same time, a dramatic reduction of the cavitation threshold is observed even at insignificant increase in the aluminium oxide content from 0.001 to 0.005%. Similar results were obtained for all aluminium alloys. Thus, cavitation treatment in the real aluminium melt can easily be implemented with the use of commercial ultrasound instruments.

4.1.2. Cavity dynamics in aluminium melt

Whilst a host of cavitation bubbles of widely different sizes (the so-called cavitation region) forms in actual melts subjected to ultrasonic treatment, it is important to describe correctly the behaviour of a single cavity in the sign-alternating ultrasonic field.

There are a great many theoretical works considering the dynamics of a single cavitation bubble, but we will apply the modified Noltingk–Neppiras equation (1950) to the description of the behaviour of a bubble in the aluminium melt, and refer the reader for a

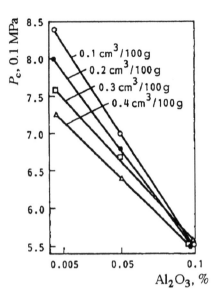

Figure 4.3   Influence of aluminum oxide and hydrogen contents in commercial A7 (1070) aluminium melt at 710°C on cavitation threshold.

detailed discussion to our original works [2–4]. In what follows, we will describe only the basic results of these studies required for understanding the essence of the homogenizing action of cavitation upon liquid metal as an alternative to HTT.

To solve the above-mentioned equation, it is necessary to set the parameters of the ultrasonic field, the physical properties of the liquid, and the initial radius of the bubble. We slightly improved these calculations, as applied to the behaviour of the bubble in the melt, by taking into account the diffusion-controlled change of the gas (hydrogen) concentration in the bubble. The results presented in Fig. 4.4 describe the pulsations of bubbles of different initial radii R_0 (from 10 to 100 μm) below the cavitation threshold and at the onset of this process. Analysis of these results points to certain differences in the evolution of the relative radius R/R_0 of the bubble and the hydrogen pressure P_g in the bubble during 2–3 periods of the sound wave in the perfect state (neglecting gas diffusion to the cavity) and in the case when the bubble, in the process of evolution of its size, is filled with gas dissolved in the melt. In both cases, as the applied sound pressure increases, the initial pulsations of the bubble follow the harmonic law but then its behavior changes radically. Upon crossing the cavitation threshold and transition to the regime of developed cavitation, the bubble rapidly expands at the half-period of rarefaction, (sometimes to ten or one hundred times above its initial dimensions), and the pressure inside it falls to deep vacuum. Then, at the compression half-period the bubble rapidly collapses.

An even more impressive picture is shown in Fig. 4.5 where curves of the evolution of bubbles with initial radius 1–100 μm are presented for two regimes of developed cavitation: with a pressure of 5 and 10 MPa. As the initial size of the bubble reduces from 100 to 1 μm, the variation of the bubble size in the developed cavitation regime obeys the general law: the smaller the initial bubble size, the more rapidly it grows and the more actively it collapses. As cavitation progresses, the distinctions in the bubble internal

198   I. G. BRODOVA *et al.*

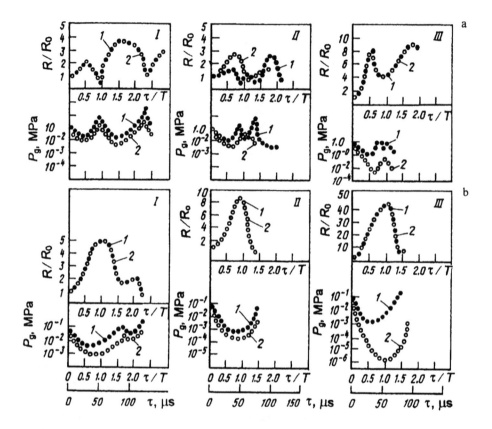

Figure 4.4   Evolution of the relative radius R/R_0 and hydrogen pressure P_g in gas bubbles (a) without cavitation (acoustic pressure $P_A = 0.2$ MPa) and (b) at the incipient cavitation ($P_A = 1.0$ MPa) for (I) $R_0 = 100$ μm, (II) $R_0 = 50$ μm, and (III) $R_0 = 10$ μm. Calculations are made with (1) and without (2) hydrogen diffusion into the bubble.

pressure calculated with and without gas diffusion grow, attaining several orders of magnitude.

Of particular interest is the study of the bubble's collapse phase. High-speed photography of adjacent cavitation bubbles in water shows that the glossy and smooth surface of the bubble changes for an indented one in the process of its expansion and after the collapse disintegrates into a multitude of tiny fragments. Taking into account the high speed of this process (tens of microseconds) and the enormous number of bubbles, we may assert that the cavitation process develops as a chain reaction. The cavitation phenomenon reproduces itself in geometric progression. This explains the efficiency of cavitation treatment of liquids and the wide-spread use of ultrasonic technology in industry.

4.1.3. Topography of cavitation field in melt

As indicated above, in actual melts a multitude of cavitation bubbles originate even at relatively low acoustic pressures insignificantly exceeding the cavitation threshold. These

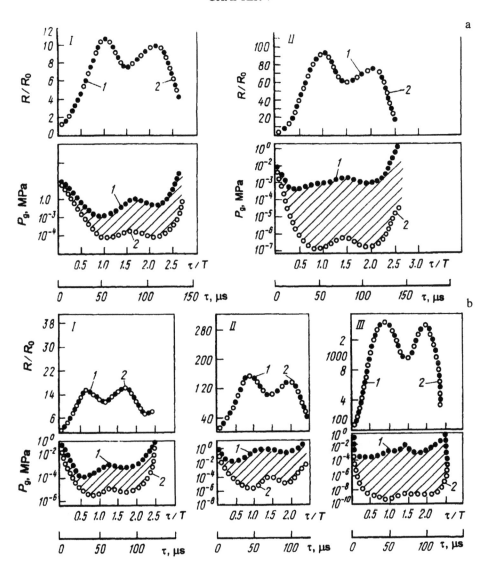

Figure 4.5   Evolution of the relative radius R/R_0 and hydrogen pressure P_g in cavitation bubbles at sound pressures of 5 MPa (a) and 10 MPa (b) in the presence of developed cavitation for (I) $R_0 = 100$ µm, (II) $R_0 = 10$ µm, and (III) $R_0 = 1$ µm. Calculations are made with (1) and without (2) hydrogen diffusion into the bubble.

bubbles occupy the part of the volume of liquid metal called the cavitation region. Its dimensions depend both on the surface of the radiating zone and on the degree of cavitation development, i.e. on the value of applied sound pressure.

The description of the topography of the cavitation region would be incomplete unless we pointed to the existence in the liquid of acoustic flows under ultrasonic treatment. This is particularly important when the effect of ultrasonic treatment on mass transfer and homogenization of melt composition is considered.

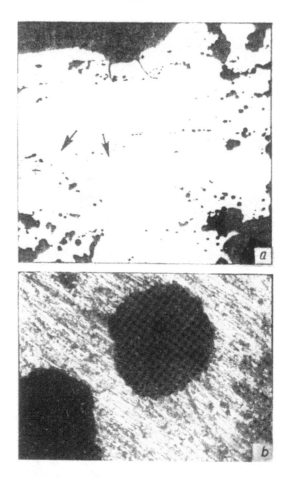

Figure 4.6   Appearance of a 100 μm titanium foil (a) with traces of cavitation erosion
(×1) and (b) the shape of a single cavitation break (×300). The foil remained in the
aluminum melt with developed cavitation for 15 min.

It is customary to distinguish the large-scale flows (their scale is determined by the sound
wavelength) and small-scale flows formed at the phase boundaries (for example, near the
surface of the oscillating cavitation bubbles). With actually achievable degree of acoustic
cavitation in melts, the velocities of acoustic streams may reach a few meters per second.
These streams play an important role in mass transfer processes and homogenization of
alloys.

An increase in the intensity of ultrasonic treatment (a growth of sound pressure) leads to
the damping of oscillations even at distances as small as a few centimeters from their
source. For this reason, the cavitation region has finite dimensions.

In general, the geometric size of the cavitation region in a transparent liquid can be
observed visually, while in melts of aluminium and its alloys they are seen indirectly, by
observing erosion of a thin foil introduced in the melt at various distances from the
ultrasound source. Typically, the linear size of this region is close to a quarter of the sound

wavelength in the given liquid. It is equal to 20–40 mm for water and 40–60 mm for aluminium melt. Figure 4.6 shows titanium foil soaked in an aluminium melt for 10–15 min under the action of ultrasonic cavitation. Punctures caused by the destructive influence of separate collapsing bubbles and traces of cavitation erosion in the form of large eroded pieces are readily seen at the foil surface.

More detailed information about the propagation of powerful ultrasound and energy losses related to cavitation is reported in [2–4].

To summarize this section, we should note once again that the development in melts of acoustic cavitation – the formation of a mass of cavitation bubbles of various initial sizes executing nonlinear oscillations in the ultrasonic field and periodically collapsing with the birth of new cavitation nuclei by the chain reaction mechanism – creates unique possibilities for melt homogenization. Below we will show that this method presents an excellent alternative to HTT owing to its high speed and the possibility of implementing it at relatively low overheating of the melt above the liquidus (no more than 50–80°C).

4.2   Effect of Ultrasonic Treatment on the Distribution of Components and Phase Constituents of Metallic Melts

4.2.1. Quick mixing of components and enhancement of melt homogeneity

In Chapter 1 we presented data attesting to the lack of mixing the components of binary and multicomponent metallic melts on the atomic level at moderate overheating of the system above the liquidus. Since such a microheterogeneous state is metastable (or non-equilibrium, with a long relaxation time), it can be irreversibly destroyed not only by heat treatment of the liquid metal but also by other energy actions. It is known that in colloid chemistry ultrasonic oscillations are used effectively for producing fine emulsions and suspensions, as well as for dissolving the components of various solutions more rapidly. For this reason, attempts to affect the degree of microheterogeneity of metallic melts by introducing ultrasound into these melts were thought to be justified.

In [7, 8] the treatment of melts by powerful ultrasonic oscillations was combined with continuous monitoring of its density by the gamma-ray absorption technique. With this aim, the authors produced a special attachment to the gamma-ray densitometer (Fig. 4.7) allowing for the introduction into evacuated melts of oscillations with a frequency of 17–35 kHz. The maximum power of these oscillations reached 2 W/cm^2, significantly exceeding the cavitation threshold for liquid metals. In the course of ultrasonic treatment (UST), the beam of gamma-ray photons irradiated continuously the area of the melt of interest, near the crucible bottom.

In the first series of experiments the authors studied the effect of UST on the duration of the relaxation process in forming a Sn–Pb alloy. To speed up mixing of the components, weighed amounts of tin and lead were loaded into a crucible in such a way that lead, which has a higher density, was placed at the top of the crucible. Upon melting the sample, the system was maintained at 370°C for 5 h, the intensity of gamma-ray flux passing through liquid metal was recorded constantly. It was found that this time is insufficient for density equilibration. In a similar experiment, equilibrium was established at the indicated temperature almost immediately after 40 s of ultrasonic treatment of the system at the

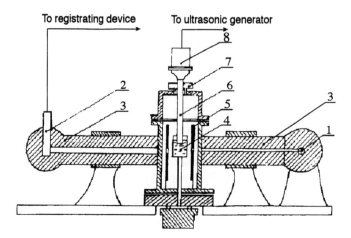

Figure 4.7   Setup for studying the effect of ultrasonic treatment on the kinetics of component mixing in metallic melts; (1) gamma-ray radiation source, (2) gamma-ray photon detector, (3) collimators, (4) specimen, (5) molybdenum heater, (6) waveguide made of titanium alloy, (7) vacuum seal, (8) magnetostrictive transducer.

maximum power of the ultrasound generator (Fig. 4.8a).

In some experiments the power of ultrasonic oscillations P was raised gradually. Simultaneously the melt surface condition was controlled visually. It was found that radical acceleration of the relaxation process was observed only above a certain threshold power, at which a host of bubbles form at the surface of the melt, which points to cavitation development. Obviously, the main contribution to the accelerated mixing of the components is made by cavitation phenomena.

Similar results were also obtained for a babbitt alloy of the Pb–Na–Ca–Al system melted in industrial conditions and subjected to ultrasonic treatment (Fig. 4.8b). In the course of isothermal holding of the melt at 500°C without ultrasonic treatment, the equilibrium density of its irradiated zone was attained only after 3 hours. The introduction of ultrasound at the maximum generator power brings the melt to equilibrium in 1 min. Upon solidification with a cooling rate of 2°C/s, the distribution of alloying elements over the ingot height was examined in the reference and sonicated specimens. It was found that the difference in calcium concentration in the top and bottom parts of an ingot of 2 cm height is 1.9%, whereas in the specimen subjected to UST this difference is only 0.16%, which means that the treatment aided the reduction of inhomogeneity if the calcium height distribution. Metallographic analysis has shown that UST changed the shape of precipitates of the intermetallic compound Pb_3Ca, which became more rounded, with blurred boundaries.

In experiments with Sn- and Pb-based alloys ultrasound was introduced in a system free of particles of the solid phase. To find out the efficiency of its influence on the rate of transition of the components from the solid state to a solution, the process of iron solution in liquid aluminium was examined at 1130°C with and without UST. An iron ingot was placed on the surface of the aluminium sample. Its mass corresponded to the Fe concentration in Al of about 8 wt. %. After the required temperature was attained, the

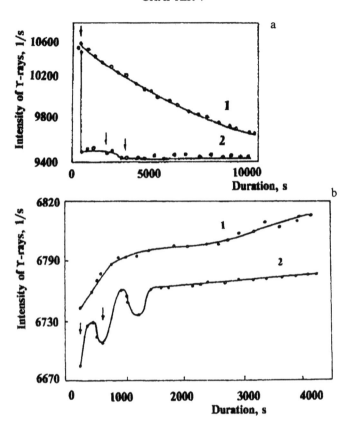

Figure 4.8   Time dependence of the intensity of gamma-ray radiation passing through (a) Pb–50% Sn melt and (b) Pb–1% Ca–0.8% Na–0.1% Al melt without (1) and with (2) ultrasonic treatment. Arrows mark the points at which ultrasound was introduced into the melt.

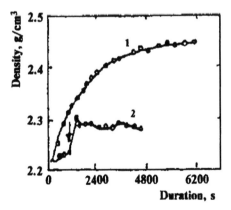

Figure 4.9   Time dependence of the density of the irradiated zone in Al–8% Fe melt obtained in the course of mixing its components at 1400°C without (1) and with (2) ultrasonic treatment. Arrow marks the time at which ultrasound was introduced into the melt.

change in density of the irradiated zone over time was measured for 90–100 min. Undissolved small iron pieces were detected in the specimens solidified 20–40 min. after the onset of isothermal holding. If the melt was subjected to ultrasonic oscillations at maximum power for 1 min. at the initial stage of solution, the relaxation was completed in 5 min. after the generator had been shut off (Fig. 4.9), which points to the effective acceleration of iron solution under the influence of UST. Metallographic analysis of the solidified samples has shown that in this case UST also promoted greater uniformity of the distribution and the change in the form and size of intermetallic inclusions.

Disagreement between equilibrium density values in sonicated and non-sonicated specimens indicates that UST not only speeds up mixing of the components, but also changes their equilibrium distribution in the system. A decrease in the density of the irradiated zone located in the bottom part of the samples, which is caused by UST, is accounted for by the more uniform height distribution of the components related to the additional dispergation or complete solution of particles of the disperse phase.

These experiments reinforced the statement made above that the treatment of melts existing in the non-equilibrium or metastable microheterogeneous state by ultrasound in the regime of developed cavitation is an effective method of homogenization of melts, which causes changes in the structure of cast metal similar to the effect of overheating. They showed that the introduction of UST in the technology of producing commercial melts may promote a considerable reduction in casting time and an improvement of service properties of products.

4.2.2. Effect of cavitation on one-way gas diffusion from liquid to bubble

Theoretical and experimental study of the dynamics of cavitation bubbles suggests that, along with its destruction at the collapse stage and formation of a multitude of new bubbles, the conservation of the bubble without collapse is also possible as a result of diffusion into the bubble of the gas dissolved in the surrounding melt.

We explain the uniqueness of the process of gas diffusion through the wall of a bubble in a liquid under the action of ultrasound [2–4, 9] by the following main circumstances. First, the surface area of a pulsating bubble in the expansion phase is much larger than at the compression phase. Therefore, the amount of gas diffusing into the bubble at the expansion stage exceeds the gas flow effluent from the bubble during the compression. Second, the diffusion process, as is known, is governed by a thin diffusion layer of liquid surrounding the bubble. At the compression stage, the thickness of this layer grows and the concentration gradient in it decreases. In contrast, at the expansion stage its thickness is reduced and the gradient increases, which causes an additional acceleration of gas supply into the bubble.

As a result, the growth of pulsating cavitation bubbles is possible. The phenomenon of a bubble's diffusive growth is called *one-way gas diffusion.*

We made an attempt to estimate the hydrogen mass diffusing into a pulsating cavitation bubble from the aluminium melt in various UST regimes. Bearing in mind certain assumptions [2–4, 6], we solved analytically the equation of the bubble dynamics and convective diffusion. Calculations were carried out for an aluminium melt with $0.2 \text{ cm}^3/100 \text{ g}$ of hydrogen using the coefficient of hydrogen diffusion in the melt $K_D = 1 \text{ cm}^2/\text{s}$. We varied the initial radius of the cavitation bubble and UST regimes (without cavitation, at the cavitation threshold, and for developed cavitation). The mass of diffused hydrogen was determined by numerical integration of the curves presented in Figs. 4.4 and 4.5, which describe the change over time of the relative radius and gas pressure in the bubble. The mean results of the calculations for some variants are summarized in Table 4.1.

Table 4.1   Hydrogen mass diffusing into gas bubble at various sound pressures and bubble's initial radii.

Sound pressure, MPa	0.2	1.0	5.0	10.0
Hydrogen mass (10^{-9} g) diffusing into bubble of initial radius:				
1.0 μm	4.0	4.6	300	2000
10 μm	0.4	4.6	300	1000
1000 μm	0.0004	0.7	200	650

It is known that the dimensions of unwettable particles of aluminium oxide actually existing in the melt and serving as cavitation nuclei do not exceed 1–3 μm, whereas the hydrogen bubbles formed at microscopic surface irregularities of these particles may have a size of 1–100 μm. The calculation results presented above show clearly that under such conditions the one-way growth of cavitation bubbles and their transformation to centres of active melt degassing are possible. The ultrasonic treatment of melt with the aim of its refinement from gaseous inclusions was one of the first successful practical applications of ultrasound in metallurgy of light alloys [2].

4.2.3. Brownian gas bubbles and effect of UST on these bubbles

Not only bubbles of cavitation origin but also the equilibrium "plankton" of finer, nanometer-sized bubbles which are permanently present in the melt volume may serve as centres for degassing gas-saturated liquids, such as molten aluminium alloys. It was thought until recently that stable equilibrium of the bubble with the surrounding melt is impossible: a gas cavity of critical size in a liquid must collapse, whereas a cavity of super-critical size will grow to macroscopic dimensions and will be removed from the system upon floating up to the surface. The source of this picture lies in the fundamental thermo-dynamical works by Gibbs, who considered the change of the free energy of a boundless liquid, upon formation in it of a nucleus of the new phase. In contrast to Gibbs' model, Popel and Kuzin [10] have recently considered the evolution of a gas bubble in a limited volume of melt. Following Rusanov's analysis, they presented the change of the Gibbs thermodynamic potential in the form

$$\Delta G = 4\pi\sigma R^2/3 + \Sigma v_i^{(1)}\mu_i^{(1)} + \Sigma v_i^{(2)}\mu_i^{(2)} - \Sigma v_i\mu_i, \qquad (4.1)$$

where σ is the surface tension coefficient, v_i is the number of moles of the ith component, and μ_i is its chemical potential. Superscripts (1) and (2) refer to the bubble and to the surrounding solution respectively, while the quantities without superscripts characterize the initial melt.

Equation (4.1) differs from the usual Gibbs expression by the fact that it takes into account the changes in the chemical potentials of the components in the surrounding solution caused by the bubble formation. This distinction is essential when the number of atoms of gas in the solution surrounding the gaseous cavity is comparable with the number of atoms in the cavity volume, i.e. in the case of nucleation of the gas phase in the limited cell of the melt. Such a situation corresponds to a sufficiently high rate of nucleation of gas nuclei, i.e. in a strongly oversaturated solution of gas in the melt. (The hydrogen content in liquid aluminium is known to surpass normally its equilibrium solubility by several orders of magnitude.) Otherwise, the change in chemical potential of the solution can be ignored and we obtain the classical dependence of the free energy on the bubble radius with the only extremum – maximum at the critical nucleus radius, which suggests an unbounded growth of bubbles of supercritical sizes.

For the binary metal-diatomic gas system, equation (4.1) takes the form

$$\Delta G = 4\pi\sigma R^2/3 + v_{g2}^{(1)}\mu_{g2}^{(1)} + v_g^{(2)}\mu_g^{(2)} + v_{me}^{(2)}\mu_{me}^{(2)} - v_g\mu_g - v_{me}\mu_{me}, \qquad (4.2)$$

where subscript "g2" refers to the diatomic gas in the bubble; subscript "g", to the gas in the solution; and subscript "me", to the metal.

In the regular solution approximation we have

$$\mu_g^{(2)} = \mu_g^0 + R_gT\ln(x_g^{(2)}) - (1 - x_g^{(2)})^2\omega_\mu;$$

$$\mu_g = \mu_g^0 + R_gT\ln(x_g) - (1 - x_g)^2\omega_\mu;$$

$$\hspace{10cm}(4.3)$$

$$\mu_{me}^{(2)} = \mu_{me}^0 + R_gT\ln(1 - x_g^{(2)}) - (x_g^{(2)})^2\omega_\mu;$$

$$\mu_{me} = \mu_{me}^0 + R_gT\ln(1 - x_g) - (x_g)^2\omega_\mu.$$

Here, x_g are the atomic fractions of the gas component in the corresponding phases, μ_i^0 are the standard chemical potentials, and ω is the interchange energy of the solution taking into account the energetically non-equivalent character of interactions of like and dislike atoms.

Assuming that the gas in the bubble is ideal, we can represent its chemical potential in the form

$$\mu_{g2}^{(1)} = \mu_{g2}^0 + R_gT\ln(p_{g2}^{(1)}), \qquad (4.4)$$

where μ_{g2}^0 is the standard chemical potential corresponding to the atmospheric pressure and p_{g2} is the gas pressure in the bubble.

Let us express the atomic fractions of the components in terms of the numbers of their moles in the corresponding phases, assuming that evaporation of metal into the bubble can be neglected, i.e. $v_{me}^{(2)} = v_{me}$, and substitute the resulting expressions in formula (4.2), which describes the change in the chemical potential related to bubble formation. Then we obtain:

$$\Delta G = 4\pi\sigma R^2/3 + v_{me}(x_g - x_g^{(2)})(\mu_{g2}^0/2 - \mu_g^0)/(1 - x_g)(1 - x_g^{(2)} +$$

$$v_{me}(x_g - x_g^{(2)})R_gT\ln(p_{g2})/(2(1 - x_g)(1 - x_g^{(2)}) +$$

$$\hspace{8cm}(4.5)$$

$$v_{me}R_gT(x_g^{(2)}\ln(x_g^{(2)})/(1 - x_g^{(2)}) - x_g\ln(x_g)/(1 - x_g)) +$$

$$v_{me}\omega(x_g - x_g^{(2)}) + v_{me}R_gT\ln((1 - x_g^{(2)})/(1 - x_g)).$$

The difference $\mu_{g2}^0/2 - \mu_g^0$ between the standard chemical potentials corresponding to the atmospheric pressure of the gas in the bubble and to the infinitely dilute gas solution in the metal is equal to the chemical potential of gas evaporation at the atmospheric pressure $\Delta G_{g-g2/2}^0$. The pressure $p_{g2}^{(1)}$ inside the bubble differs from the external pressure p by the value of the capillary constituent $2\sigma/R$, where R is the bubble radius:

$$p_{g2}^{(1)} = p + 2\sigma/R. \hspace{4cm}(4.6)$$

Substituting these expressions into equality (4.5) gives the final expression for the change in the thermodynamical potential of the system caused by bubble formation:

$$\Delta G = 4\pi\sigma R^2/3 + v_{me}(x_g - x_{g2}^{(2)})\Delta G_{g-g2/2}^0/((1 - x_g)(1 - x_g^{(2)})) +$$

$$v_{me}(x_g - x_g^{(2)})R_gT\ln(P + 2\sigma/r)/(2(1 - x_g)(1 - x_g^{(2)})) +$$

$$\hspace{8cm}(4.7)$$

$$v_{me}R_gT(x_g^{(2)}\ln(x_g^{(2)})/(1 - x_g^{(2)}) - x_g\ln(x_g)/(1 - x_g)) +$$

$$v_{me}\omega_\mu(x_g - x_g^{(2)}) + v_{me}R_gT\ln((1 - x_g^{(2)})/(1 - x_g)).$$

The number of moles of gas in the bubble can be related to the pressure inside the bubble, the temperature, and the bubble volume, using Clapeyron's equation (the gas in the bubble, as we mentioned above, is assumed to be perfect). The gas concentration in the melt surrounding the bubble, which enters Eq. (4.7), can then be represented in the form:

$$x_g^{(2)} = (x_g - 8\pi(pR^3 + 2\sigma R^2)(1 - x_g)/(3v_{me}R_gT))/(1 - 8\pi(pR^3 + 2\sigma R^2)(1 - x_g)/(3v_{me}R_gT). \hspace{0.5cm}(4.8)$$

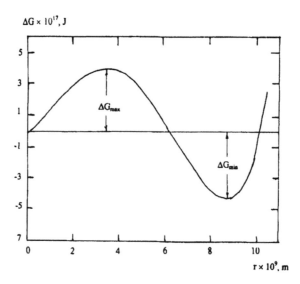

Figure 4.10   Dependence of the change ΔG of the Gibbs free energy caused by bubble formation on bubble radius ($v_{me} = 3.5 \cdot 10^{-14}$ mol, $x_g = 1 \cdot 10^{-5}$, $p = 5 \cdot 10^4$ Pa, $T = 700°C$).

Thereafter one is left with the following independent parameters determining the state of the system under consideration: the temperature T, pressure p, surface tension σ at the bubble boundary, the system's size defined by the number v_{me} of moles of metal, and the initial gas concentration in the melt x_g. The "individual marker" of the system in expression (4.7) is the interchange energy ω, which can be found from the experimental heat of mixing, saturated vapour pressure, or by other methods.

Let us examine the behaviour of the quantity ΔG as a function of various parameters. Calculations will be carried out for aluminium–hydrogen solutions. The interchange energy of this system is -38.5 kJ/mole. The surface tension at the bubble boundary is assumed to be 0.914 mJ/m^2, which fits the experimental data on the surface tension of liquid aluminium at the boundary with a vacuum. The size of the system is chosen so as to make the volume fraction of the gas phase in the system equal, in order of magnitude, to the volume fraction of the vacancy subsystem in liquid aluminium at the given temperature.

The typical dependence of the change ΔG of thermodynamical potential related to bubble formation on the radius of the bubble is presented in Fig. 4.10. Unlike the usual Gibbs dependence with one maximum, in this case the $\Delta G(R)$ curve has also a minimum corresponding to the metastable or stable equilibrium of the bubble with the surrounding melt.

Thus, we confirmed thermodynamically the possibility of long conservation in a liquid metal of equilibrium nanometer-sized gas bubbles and, therefore, the possibility of long existence of a gas-saturated melt in a state of thermodynamically stable or metastable foam.

At sufficiently high hydrogen content x_H in the system, the quantity ΔG_{min} corresponding to the minimum of the $\Delta G(R)$ curve is negative (Fig. 4.11). Consequently, in this case gas bubbles are thermodynamically stable. As x_H decreases, the system changes over into the states with $\Delta G_{min} > 0$ and the equilibrium of the bubble with the surrounding melt becomes metastable. It can be violated by imparting the activation energy $\Delta G_{max} - \Delta G_{min}$ to the

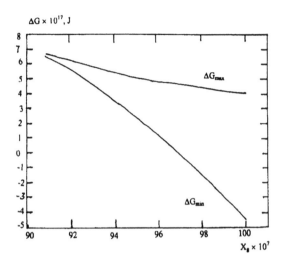

Figure 4.11   Dependence of ΔG_{min} and ΔG_{max} on the initial gas concentration x_g in the system ($v_{me} = 3.5 \cdot 10^{-14}$ mol, $x_g = 1 \cdot 10^{-5}$, $P = 5 \cdot 10^4$ Pa, $T = 700°C$).

system (here ΔG_{max} is the value of the function $\Delta G(R)$ corresponding to the maximum of the curve in Fig. 4.11) and then the bubble dissolves in the melt. Clearly, the hydrogen concentration at which ΔG_{min} changes sign corresponds to the limiting hydrogen solubility in the given volume of liquid aluminium. With a further reduction in the gas content, the stability of the metastable state declines and, finally, at certain x_H^* value, extrema in the $\Delta G(R)$ curve disappear and it takes a parabolic form concave to the horizontal axis. Below this concentration the existence of metastable gas bubbles in the melt becomes impossible, and the point x_H^* can be considered as an analog of the spinodal point for the colloidal state of liquid metal.

Colloidal states of the gas-saturated solution cannot be realized in all systems, but only when there exists rather strong predominance of the interaction of metal atoms with gas atoms over the interaction of like atoms, i.e. at rather strong negative values of the interchange energy ω. As ω approaches a certain threshold value ω_μ^*, the stability of the equilibrium of the bubble with the surrounding melt decreases. However, the absolute value of interchange energy for the aluminium–hydrogen system exceeds significantly the critical value ω^* and we have every reason to suggest that a significant portion of the hydrogen occluded by aluminium melts is confined precisely in the equilibrium bubbles. Owing to the small dimensions, these bubbles do not float up to the surface and move within the volume of liquid metal like Brownian particles.

In the field of ultrasonic oscillations these bubbles begin to oscillate, gaining additional energy. If the latter is sufficient to overcome the energy barrier ($\Delta G_{max} - \Delta G_{min}$) separating the equilibrium bubble from the state of true solution, the bubbles will dissolve. The solution may prove to be either thermodynamically stable (if $\Delta G_{min} > 0$) or metastable, oversaturated ($\Delta G < 0$). In the latter case, after a sufficient time has lapsed or under the influence of external factors (increase in temperature, a second UST, etc.) the solution repeatedly goes over into the state of ultrafine foam. Moreover, the process of bubble merging or floating up to the melt surface can be initiated by ultrasonic oscillations.

Thus, the treatment of a gas-saturated metallic melt by ultrasound appears to be a quite promising method of action upon the system of Brownian nanometer-sized gaseous bubbles. It is possible that its influence would be even more effective in the precavitation mode. However, special experiments on the influence of ultrasonic oscillations on the behaviour of gas "plankton" have not been carried out to date. The existence of such microscopic bubbles in the melt is confirmed by the presence in aluminium ingots of a great number of pores with a characteristic size of 10–100 nm. We can assume that these pores are inherited from the initial melt. Porosity on the indicated scale accounts for a significant part of the overall porosity of cast metal. As a consequence, varying the content of gas "plankton" in the melt by means of properly chosen UST regimes enables us to control such an important characteristic of cast metal as its porosity.

4.2.4. Activation of non-metallic inclusions as solidification nuclei

It is known that in metallurgical installations the solidification process proceeds by a heterogeneous mechanism. The main controlling factors of heterogeneous solidification are the melt temperature, the melt purity with respect to active impurities as solidification centres, and the rate of cooling of the melt in the solidification process. As has been already mentioned, the UST of melt in the cavitation mode promotes wetting of non-controllable non-metallic impurities and introduces them in the solidification process as active supports for forming crystals of transition metal (Ti, Zr, etc.) aluminides and other primary phases.

We will not discuss here the action of ultrasound on the solidification front. According to the conclusions of the theory [11], in this case cast grain refinement occurs as a result of destruction of growing dendrites and formation of new solidification centres ahead of this front in the form of suspended fragments of dendritic branches.

When treating the melt by ultrasound in the cavitation mode, we change the structure of liquid metal far away from the solidification front. In this case the creation of excess solidification centres ahead of the solidification front, provided by the involvement of a host of nuclei activated by cavitation in the solidification process, is a powerful factor of ultrasound's influence on the structure of liquid metal and the solidifying ingot.

In our works [2–4] we develop the ideas of famous Russian scientists in the field of physical metallurgy (Baikov, Danilov, Kazachkovskii and others) on the significant role of non-metallic impurities in the solidification process. Danilov and Kazachkovskii established, in particular, that the process of transformation of the impurity into an active solidification centre depends to a great measure on the surface relief of impurity particles (the shape and size of fissures and cracks at this surface) and on the interphase tension at the impurity–melt boundary. The impurity is activated as a result of penetration into the fissure or crack and the initiation in it of the first crystals of the alloy. These microscopic crystals may persist in these cavities even on heating above the liquidus temperature of the given alloy. The *impurity activation process* takes considerable time and speeds up as the melt temperature increases.

As a consequence, the activation of fine non-metallic inclusions (as a rule, oxides of the major components), which always exist in the real melt, is achieved in the process of HTT of the melt. After such treatment, the number of active impurities increases, which is one of the main reasons for refinement of the ingot structure.

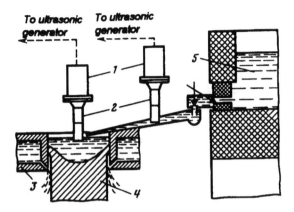

Figure 4.12   Schematic view of ultrasonic treatment of melts in processes of continuous (a) and shaped (b) casting; (1) transducer, (2) oscillating system, (3) mould, (4) mixer with melt.

Ultrasonic treatment in a cavitation regime makes it possible to quickly activate the impurity inclusions and does not require an increase in the melt temperature. Our understanding of the mechanism of activation of unwettable impurities in the cavitation field is directly related to the dynamics of cavitation bubbles. We consider unwettable fine impurities in the melt as a system consisting of the melt, gas, and particle, because the presence of a fissure in the microrelief of this particle determines the existence of gas in it. Such an unwettable particle with the gas adsorbed in microscopic irregularities of its surface is a nucleus for the formation of the cavitation bubble. (We should recall Frenkel's statement [1] that the wettability phenomenon, i.e. the presence of the gas phase at least at part of the particle surface, rather than the size of unwettable particles in the liquid, determines the possibility of their transformation into cavitation nuclei.) Cavitation bubbles are reproduced by means of a chain reaction, when a single bubble decays after 2–3 pulsations into several new fragments and so on. However, a certain part of them collapse generating a high pressure pulse. Such pulses promote wetting, i.e. the filling of microcracks existing at the surface of non-metallic particles by the matrix melt. Having penetrated into the crack, the melt readily solidifies upon slight supercooling. Subsequently, it may retain the crystal structure even above the liquidus temperatures and on subsequent supercooling of the melt stimulates solidification of the remaining system [2–4].

4.3   Effect of Ultrasonic Treatment of Melts on the Ingot Structure

4.3.1. Basic arrangements for ultrasonic treatment of melts

Irrespective of the goals of using ultrasonic treatment, be it degassing of the melt and filtering applied to a mould, or continuous casting of light alloys, or direct impact on the solidification process, ultrasound is always introduced in liquid metal.

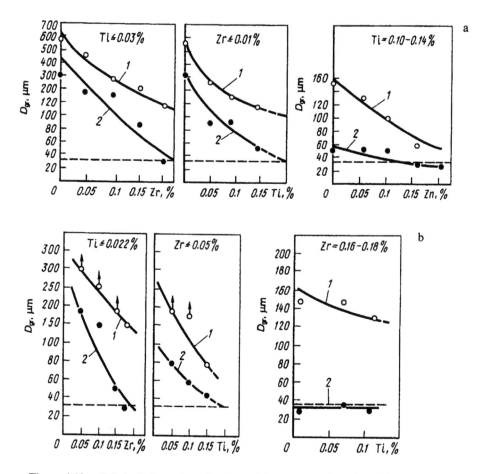

Figure 4.13   Effect of ultrasonic treatment and the concentration of modifying additives on grain refinement in 70 mm diameter ingots of (a) D16ch (2324) and (b) V95pch (7475) alloys: casting without (1) and with (2) sonication. Dashed lines show the dendritic cell size in ingots cast without sonication.

The principal schemes for ultrasonic treatment, as applied to continuous casting of ingots of aluminium alloys, are shown in Fig. 4.12. Common for these schemes is the acoustic independence of the crystallization front on the ultrasound source. When treating the melt en route to the mould or to the mixer, the refining possibilities of ultrasound – degassing or filtering metal from solid and non-metallic impurities – are normally used. In the latter case a multilayer screen filter is placed en route to the mould [2–4]. Below we will show that along with melt degassing, cavitation treatment assists in grain refining of the metal structure. In this case, *the activating effect of ultrasonic treatment on non-metallic inclusions contained in liquid metal*, which was discussed in the previous section, is of paramount importance. In the case of very fine filtering of the melt in an ultrasonic field, while retaining disperse inclusions of size smaller than 1 μm in the filter, the impact of ultrasound on the process of refinement of the ingot structure weakens or entirely disappears. This is accounted for by the fact that the filter traps non-metallic particles activated in the field of acoustic cavitation, which could play the role of solidification nuclei.

The foregoing can be illustrated by the curves showing the dependence of refinement of the structure (Figs. 4.13a,b) of ingots of 70 mm diameter, prepared from alloys D16ch (2324) and V95 pch (7475) with Zr and Ti additives, on the degree of modification and the action of cavitation treatment (curve 1), as well as by the curve obtained without ultrasonic treatment (curve 2). In parentheses we indicate the alloy grades according to the USA standard.

The optimal combination of modification (the overall content of Ti and Zr being 0.20%) and UST leads to the maximum possible refinement of the grain structure of the ingot and to the formation of a so-called non-dendritic structure in which the grain size is equal to the dendritic parameter. Casting of similar ingots without UST, even with a degree of modification exceeding 0.5%, does not change the dendritic character of the structure. When the content of modifiers reaches 0.2%, we find in this structure large isolated aluminides of titanium (in the form of rosettes) and zirconium (in the form of plates).

The scheme of ultrasonic treatment of melts in which the ultrasound source produces a cavitation field directly near the crystallization front (Fig. 4.12) is of great importance for limiting grain refinement of an ingot and formation of a non-dendritic structure. In this case the conversion of acoustic energy to heat causes significant overheating in the liquid bath of the ingot above the liquidus temperature and reduces the distance for transporting a great number of crystallization centres into the zone of overheating near the crystallization front.

Much can be learned from the results of comparative studies of casting of ingots of 270 mm in diameter prepared from the 1960 (7055) alloy of the Al–Zn–Mg–Cu system with 0.16% Zr in which two technological UST schemes were employed (Fig. 4.12). When the treatment was carried out in the melt flow en route between the mixer and mould, the melt temperature exceeded the liquidus temperature of the alloy by 80–100°C, while in the case of UST performed in the liquid bath of the ingot it was higher than the liquidus temperature by only 10–15°C. Results showing the different degree of refinement of grain structure and the reduction in the hydrogen content (the degree of melt degassing) under the action of ultrasound are presented in Table 4.2.

Table 4.2 Effect of UST site on the efficiency of grain refinement and reduction of hydrogen content in 270 mm diameter ingots of 1960 (7055) alloy.

Ultrasonic treatment (UST)	Temperature, °C	Number of ultrasound sources	Grain size, μm	Hydrogen content, cm³/100 g
No UST	720	–	< 1000	0.24
UST en route to mould	720	1	200	0.07
UST in liquid pool	650	1	60*	0.12
	650	3	50*	0.07

*non-dendritic structure

Analysis of the data in this table suggests that the degree of grain refinement depends on the site where the UST is applied and, correspondingly, on the temperature conditions of cavitation treatment, activity of impurities, and the length of the path of active nuclei to the

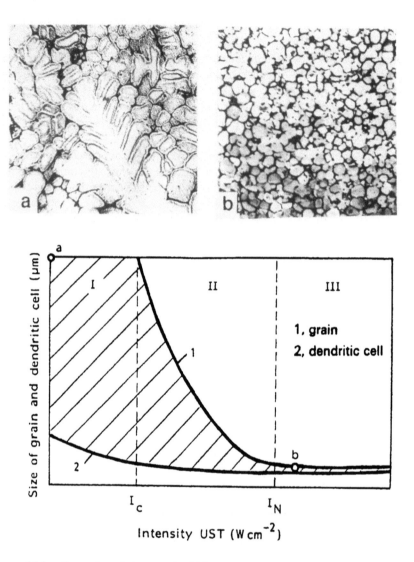

Figure 4.14 Formation of the non-dendritic structure in ingots of alloy 1960 (7055) versus the intensity of ultrasonic treatment. The photographs (a, b) represent the characteristic microstructure (×100) of the ingot. Region I at the bottom panel corresponds to the absence of cavitation; the region II, to the cavitation threshold; and the region III, to the developed cavitation mode. (1) is the grain size and (2) is the dendritic parameter (the size of dendritic cell).

crystallization front. The best grain refinement and the formation of a non-dendritic structure is achieved when UST is applied to the melt in the liquid bath of an ingot. In contrast, the efficiency of degassing at the same acoustic power (the same number of oscillation sources) proves to be higher when the melt flow is treated by ultrasound en route from the mixer to the mould.

4.3.2. Mechanism of non-dendritic solidification in ultrasonic field

Studying the effect of ultrasound on solidification of light alloys [2–4] allowed us to find out conditions for the formation of a new structure type with grain size equal to the size of the dendritic cell at the given rate of melt cooling. Such conditions suggest the combined impact on the melt of nucleation-type modifiers and ultrasonic treatment in the cavitation mode. Ultrasound energy introduced into the liquid pool of an ingot provides overheating of the melt and produces an additional temperature gradient near the crystallization front, while the developed cavitation activates impurity particles suspended in the melt and creates an excess of crystallization centres near the crystallization front.

The parameters of the structure of a 70 mm diameter ingot of 1960 (7055) alloy at various degrees of cavitation development (UST intensity) are presented in Fig. 4.14a. It is easy to separate three characteristic treatment regimes in these curves: regime I prior to the cavitation threshold, regime II in which cavitation processes originate and their ordering occurs, and regime III of developed cavitation. It is clear that in the latter regime the grain refinement curve crosses the curve showing the change in the dendritic parameter (cross-section of dendritic branches), which determines the formation of a non-dendritic ingot structure. Typical structures of ingots of this alloy – dendritic (in the initial state) and non-dendritic (as a result of acoustic cavitation) – are shown clearly in Fig. 4.14b.

Non-dendritic grains may form only if the smooth surface of the crystallization front remains stable throughout the solidification period (from the appearance of nuclei to the disappearance of the last portions of liquid metal). This is possible only at a very low melt cooling rate near the surface of the growing grain. Thus, the absence of a dendritic grain structure indicates that the growth of grains occurred in a weakly supercooled melt. The melt supercooling is caused by the need for energy expenditure for forming solidification nuclei. With the great number of modifier particles and non-metallic inclusions activated by cavitation and facilitating the formation of crystalline nuclei, supercooling at the solidification front decreases considerably.

The mechanism of formation of non-dendritic grains indicated above is confirmed by the results of the study of dendritic segregation by EPMA technique [2–4]. The lines of equal copper concentration in the non-dendritic and dendritic structures of 70 mm ingots of 1161 (2324) alloy are presented in Fig. 4.15. It is evident that a non-dendritic grain grows uniformly in all directions by increasing the grain radius, without partition into dendritic branches. As a non-dendritic grain develops, its shape may deviate slightly from the spherical one on account of the counter growth of other non-dendritic grains (Fig. 4.15a). Similar curves constructed for a large dendritic branch (Fig. 4.15b) indicate that it also grows in all directions. However, in forming a non-dendritic structure and spherical grains, the alloying component accumulates less intensively than in the case of the growth of a cylinder that may be considered as a model of a dendritic branch. *This finding illustrates the effect of acoustic cavitation not only on the structural but also on the chemical homogeneity of the ingot.* It is possible that the increase in the latter is related to a great extent to the homogenizing effect of UST on the initial melt.

It should be recalled that the formation of non-dendritic structure in particular solidification conditions was observed earlier for various alloys. For example, Flemings [11] obtained a similar structure on solidifying an alloy of the Mg–Zn–Zr system in a metal

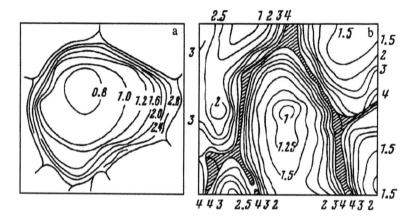

Figure 4.15   Lines of constant copper concentration in (a) non-dendritic and (b) dendritic structures of a 70 mm diameter D16ch (2324) alloy ingot in the case of casting (a) with UST (non-dendritic structure) and (b) without UST (dendritic structure). The shaded areas mark the eutectic along the dendritic cell and non-dendritic grain boundaries.

mould without external actions, as a result of the strong modifying influence of zirconium. In the same work, a coarse non-dendritic grain of about 1 mm size was produced in an Al–Cu–Ti alloy after 20-h treatment of the melt by vibration at a temperature close to the liquidus point. A non-dendritic structure in ingots of medium diameter (smaller than 300 mm) was obtained without UST in aluminium alloys with scandium addition, which is one of the strongest nucleation-type modifiers [12]. Finally, in nickel alloys such a structure was obtained after rapid solidification, in which case strong supercooling of the melt is produced ahead of the crystallization front by an alternative solidification mechanism (homogeneous nucleation) [13, 14].

The use of UST in the cavitation mode allowed the non-dendritic structure to be obtained on an industrial scale for most aluminium alloys containing modifying additives of nucleation type, without limitations on size and shape of ingots (rounded, from 65 to 1200 mm in diameter, and flat, from 90×250 mm to 400×1200 mm in cross-section).

Our studies of non-dendritic solidification permitted us to formulate general principles of forming such a structure both for slowly solidifying objects (precise casting) and for relatively rapid solidification at continuous casting of 65–1200 mm diameter ingots and, finally, for granules and flakes obtained by ultrarapid solidification. At higher cooling rates of the melt and therefore with a higher degree of overheating at the crystallization front, a great number of active nuclei of the new phase is necessary. It is these possibilities that UST provides in the cavitation mode.

With an excessive number of solidification nuclei, *the cooling rate becomes a factor controlling the grain size in ingots (castings).* The generalized diagram illustrating this new aspect of the solidification theory is presented in Fig. 4.16. We show on it the size of non-dendritic grains in aluminium and magnesium alloys as a function of cooling rate which varies over a wide range. Analysis of these data shows clearly that all the results concerning the formation of grain (non-dendritic) structure fall within a narrow strip close to the known dependence of dendritic parameter on cooling rate.

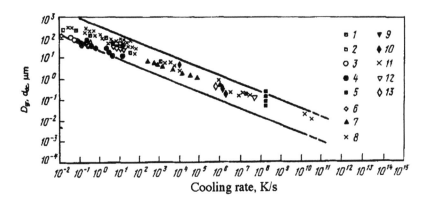

Figure 4.16   Effect of cooling rate of melt on (11–13) non-dendritic grain size and (1–10) dendritic parameter for (1–8, 11) aluminum, (9, 12) magnesium, and (10, 13) nickel alloys according to the authors' studies [2–4] and taken from [13, 14].

This new principle of the solidification theory, suggesting that the size of the non-dendritic grain and the distance between dendrite branches (dendritic parameter) are determined by the same factor – the cooling rate – was recognized to be a scientific discovery (diplome no. 271, 1983) [15, 16].

4.3.3. Advantages of non-dendritic ingot structure

The formation of a non-dendritic structure in an ingot increases homogeneity of the chemical composition over the ingot cross section. Zonal segregation in ingots with non-dendritic structure is virtually absent, which has a positive effect on the technology of subsequent treatment of cast and deformed metal. Another considerable advantage of non-dendritic ingot structure is a rise in the plasticity margin both at room temperature and in the temperature range of deformation (Figs. 4.17, 4.18). To estimate the degree of technological plasticity in the range of temperatures of plastic deformation (Fig. 4.18), we used the parameter B, the ratio of impact toughness of a metal to its yield strength. It should be noted that this advantage of non-dendritic structure is extremely important for the technology of production of high-strength structural aluminium alloys. In semi-finished products made of these alloys the rise of strength characteristics is usually achieved with ease by applying special heat treatment, but in this case it is difficult to retain acceptable plasticity. The formation of ingots with non-dendritic structure provides increased placticity of metal not only in the as-cast condition but also at all subsequent technological conversions, including final heat treatment conducted with the aim of achieving maximum mechanical properties.

It should also be noted that the rise in plasticity caused by non-dendritic solidification increases the resistance to crack formation in casting large-sized (650–1200 mm in diameter) ingots on account of the decrease in size and number of surface cold shuts. For example, for 830–960 mm diameter ingots made of 1173 (7475) alloy the depth of cold shuts decreased from 52–56 mm to 10–30 mm [2, 4]. Another important result of improving the surface quality is the decrease in losses caused by machining of the metal.

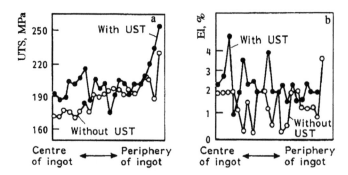

Figure 4.17   Effect of non-dendritic structure on mechanical properties (ultimate tensile strength (a) and relative elongation (b)) in the radial direction of annealed large ingots (830 mm diameter) of alloy 1161 (2324) at 20°C.

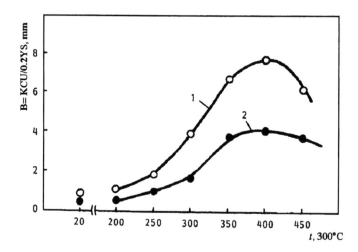

Figure 4.18   Effect of non-dendritic structure on the parameter of technological plasticity, B (ratio of impact toughness to yield strength), in the temperature range of plastic deformation of large (830 mm diameter) ingots of alloy 1973 (7475); (1) non-dendritic structure, (2) dendritic structure. The properties were determined in the ingot longitudinal direction at ingot halfradius.

A similar, but weaker effect is provided by UST of melt en route to the mould, when a non-dendritic structure does not arise but a finely differentiated dendritic structure forms.

4.3.4. Effect of non-dendritic structure on deformation in the solid–liquid state

Close attention of physical metallurgists has recently been focused on the behavior of cast metal with non-dendritic structure in the process of deformation in the solid–liquid state. Studies performed by Flemings [17] showed that the loads which must be applied to cast metal in the solidification range depend substantially on its structure. For the same degree of deformation (for instance, upsetting) billets with dendritic structure require an effort that is by an order of magnitude greater than billets with globular (non-dendritic) structure. Since it is rather difficult to obtain an ingot with non-dendritic structure, a fine dendritic structure is usually obtained in the solidification process by means of various mixing procedures. Then, on holding in the solid–liquid state, this fine dendritic structure converts into a coarse non-dendritic structure exhibiting so-called *thixotropic* properties. Thixotropy manifests itself in a drastic decline of toughness after applying force to a billet in the solidification range. By analogy with deformation in the solid state of fine-grained alloys, we can speak of the *superplasticity* of a metal with non-dendritic structure under its deformation in the solid–liquid state.

Currently this deformation method, with the volume fraction of the solid phase within 60–90%, finds extremely wide application in producing stampings of different purpose. As a rule, middle-strength alloys (A356–A357 type) of the Al–Si–Mg system are used in their production, while electromagnetic mixing of the melt in a liquid pool mould is employed for ingot grain refinement.

Experience in forming the non-dendritic structure of ingots with the use of ultrasonic (cavitation) treatment of melts allows us to propose a new technical solution for solid–liquid deformation based on the formation of the finest (for the given ingot cross section) non-dendritic structure immediately in the casting process. In this case we remove from the currently employed technological scheme the operation for converting the fine dendritic structure into a coarse non-dendritic structure in the process of special heating, and the billet with the finest non-dendritic grains used for deformation can be deformed with application of lesser loads.

The properties of as-cast billets with dendritic and non-dendritic structures in the solid–liquid state have not yet been studied in detail. We will set forth, therefore, the results of our work [18] devoted to the study of thixotropic properties of ingots of one of the highest-strength aluminium alloys, grade 1960 (7055), with non-dendritic structure. These data do not cover all the thixotropy characteristics: we used a simple method of specimen upsetting, whereas according to Flemings data [17] the maximum thixotropic effect is observed in shear deformations.

We examined samples of 285 mm diameter ingots of 1960 (7055) alloy with dendritic (without UST) and non-dendritic structure (with UST) obtained under industrial conditions. The size of non-dendritic grains in ingots of this diameter was 60–70 µm, while in ingots with dendritic structure it was 1.5–2.0 mm.

For tests in the solid–liquid state we cut out rectangular $35 \times 35 \times 12$ mm billets from cross templates of ingots with both structure types. The specimens were subjected to upsetting with a twofold decrease in height in smooth block heads on a hydraulic press under isothermal conditions (on heating and holding for 15 min). For comparison, we also carried out the same experiments on similar specimens in the solid state. Upsetting was carried out at temperatures of 420, 525, 550, 570, and 590°C. The amount and distribution of the liquid phase was estimated from the specimen microstructure after quenching from the deformation temperature. Table 4.3 contains the results of quantitative processing of tests performed with these specimens.

As follows from these results, the maximum force (and maximum upsetting stress) for ingots with non-dendritic structure proved to be much weaker than analogous values used for deformation of ingots with dendritic structure. The higher the content of the liquid phase

Table 4.3   Effect of the billet structure on the value of true stresses in 50% upsetting in the solid (at 420°C) and solid–liquid states with a rate of 0.1 mm/s.

Temperature, °C	True stress, MPa	
	Non-dendritic structure	Dendritic structure
420	103.4	111.1
525	35.8	50.4
550	21.4	30.5
570	19.3	30.9
590	11.7	19.6

in metal, the larger this difference is. It is also worth noting that the effect of the structure type on upsetting in the solid state was much lower than that in the two-phase state.

The assessment of mechanical properties in tensile tests of billets deformed by upsetting in the solid–liquid state after standard heat treatment did not reveal a reduction in strength and plasticity properties (Table 4.4). Moreover, these characteristics increased somewhat and satisfied the standard requirements for small stampings obtained by deformation in the solid state.

Table 4.4   Effect of the billet structure type on mechanical properties of upset test billets of 1960 (7055) alloy at 570°C.

Structure type	Mechanical properties	
	Ultimate strength, MPa	Elongation, %
Non-dendritic	671	10.9
Dendritic	650	7.8

Experimental data show with certainty that the use of ingots with non-dendritic structure for deformation in the solid–liquid state offers a series of advantages over the use of ingots with dendritic structure. In Flemings' opinion [17], the most important advantage may be the possibility of producing large semi-finished products for which the non-dendritic structure of the initial ingot has a crucial significance. It is this possibility that is offered by UST technology at continuous ingot casting.

4.4   Effect of Cavitation Melt Treatment on Primary Crystallization of Natural Aluminium-based Composites

In the preceding sections we considered the principles of successive solidification of aluminium alloys in an ultrasonic field with formation of the grain structure of a solid solution. In this case the appearance of primary crystals in the volume of a liquid metal is undesirable because of the formation of defects in the as-cast structure. Such crystals representing, as a rule, intermetallic compounds, form in the melt en route to the mould (at decreased casting temperatures and with excess transition metal in the alloy), or more frequently in the volume of the liquid pool. They are then trapped by the crystallization front and penetrate into the solidified part of the ingot. When designing new compositions of aluminium alloys with increased concentration of high-melting transition metals (Zr, Sc, Ti, etc.), a search for effective methods for preventing the occurrence of large-sized aluminide inclusions in the ingot structure is required.

Special aluminium alloys – natural composites in which suspended crystals of the excess phase is the normal structural constituent – are a special case. Hypereutectic silumins, the use of which in engineering is continuously growing, fall in this category [19].

According to Dobatkin [20], one of the main ways of influencing the grain refinement of suspended crystals of intermetallic compounds is the introduction of activated additives. It is assumed that at very large numbers of nuclei of a new phase near the crystallization front, the amount of "construction material" in the melt is insufficient for the growth of suspended crystals to significant dimensions.

From this point of view, UST of melt in cavitation mode may become the most acceptable way of controlling the structure of natural composites. The principles of UST influence on the solidification of intermetallics in aluminium alloys were considered by us in more detail in [2, 4]. In this section we will concentrate on the solidification process in an ultrasonic field of hypereutectic silumins containing from 17 to 25% Si.

These alloys are now widely used in practice of shaped casting and liquid stamping for producing internal-combustion engines. A low linear expansion coefficient, in combination with high wear resistance, determines the enhanced performance of pistons made of hypereutectic silumins. It is worth noting that primary silicon crystals formed in silumins in shaped casting and liquid stamping have sufficiently large dimensions (not less than 60–120 μm). This enables shaped items to be obtained but rules out the possibility of continuous ingot casting and the fabrication of deformed semi-finished products by using the classical ingot–deformed semi-finished product scheme.

At the same time, the use of continuous ingot casting technology and subsequent pressing or stamping of semi-finished products, which is typical of metallurgical practice, appears to be very promising from an economical standpoint. For this purpose, the size of primary silicon crystals must be lowered by at least a factor of 2–4 lest the formation of the normal crystallization front in continuous casting be impeded and the subsequent hot deformation process be complicated.

Such a possibility is offered by UST of liquid silumin en route to the continuous casting mould. Especially good results can be obtained with UST coupled with modification with small (up to 0.01%) phosphorus additions. The use of ultrasound enables one to readily decrease the size of primary silicon crystals in 65–170 mm ingots to 20–30 μm, which allows their extrusion and stamping to obtain various kinds of semi-finished products (Fig. 4.19).

Of special interest is the technology of producing ingots of hypereutectic silumins in which inexpensive charge (Al–30–40% Si) obtained by the electrothermal method is used as a starting material [19]. Figure 4.20 shows the microstructures of charge base (electrothermal hypoeutectoid silumins with 36% Si) and 145 mm diameter ingots of 01390 alloy with 18% Si obtained without modification and after modification with the use of combined technology.

The study of properties of extruded semi-finished products made of 01390 alloy obtained with this technology shows that they compare well with the properties of AD31–AD33 (6063, 6061) alloys, which are most widely used for producing construction materials.

Figure 4.19   Hot extruded and stamped semi-finished products obtained from ingots of hypereutectic silumins with grain structure modified with the aid of ultrasonic treatment.

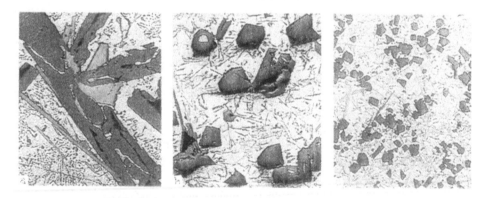

Figure 4.20   Microstructure ($\times 100$) of hypereutectic silumin: (a) charge of electrothermic silumin with 36% Si; (b) unmodified 145 mm diameter ingot of alloy 01390 with 18% Si; (c) same, after complex modification (0.01% P and UST).

However, in contrast to the latter the 01390 alloy has a 25% lower linear expansion coefficient and 10–15% higher elasticity modulus. Deformed semi-finished products prepared from this alloy are easily welded and have high corrosion-resistant properties.

4.5   Technological Aspects of Ultrasonic Treatment of Liquid Aluminium Alloys

4.5.1. Continuous casting

We have already pointed out that UST with continuous casting (Fig. 4.12) can facilitate various technological processes (degassing, fine filtering), as well as the formation of the refined structure of natural composites and non-dendritic alloy solidification.

For example, the use of UST of melt flow en route from the mixer to the mould with the goal of degassing in the continuous casting of large-sized flat ingots of AMG6 (6% Mg) alloy with 300×1700 mm cross-section, made it possible to reduce the hydrogen content from 0.6 to 0.3 $cm^3/100$ g, which is equivalent to vacuum treatment of the melt in the mixer for 1.0 h [2–4]. Below we will describe the effect of this treatment on the properties of ingots and deformed semi-finished products made of AMG6 alloy, which are widely used in welded constructions of critical purpose.

Fine filtering of the melt through multilayer screen filters enables one to improve the purity of ingots with respect to oxide inclusions by nearly one order of magnitude, which is of critical importance for the quality of stamped and pressed semi-finished products made of structural aluminium alloys used for the production of loaded parts of aerospace constructions.

Industrial use of UST with the aim of forming a non-dendritic structure made it possible to determine the optimized parameters for casting and UST (Table 4.5). When modifying alloys with optimal additions of transition metals, the casting parameters listed in this table provide formation of a non-dendritic structure in most aluminium alloys.

Figure 4.21 shows the appearance of one of the largest ingots (10 tons, diameter 1200 mm) with non-dendritic structure, obtained from 1161 (2324) alloy. The non-dendritic structure in ingots of such large dimensions was obtained as a result of ultrasonic treatment of the liquid pool of the ingot with the use of 9 ultrasound sources [2, 4].

Additional effects can be achieved by combining UST in a liquid pool with other present-day methods of continuous casting. This refers to the process of casting into an electromagnetic mould and casting into a hot top mould. According to our data [4], the use of UST with casting into an electromagnetic mould makes it possible to avoid the state with different grain sizes over the ingot cross-section and obtain, with optimal modification, a non-dendritic grain.

The use of UST in continuous casting into the hot top mould by the Air–Slip method, when solidification occurs in a cooled graphite ring with increased solidification rate, is very promising.

4.5.2. Shaped casting of aluminium alloys with the use of ultrasonic treatment

It is well known that in the shaped casting processes (especially in precision casting into ceramic and gypsum moulds) the structure is coarsened due to the slow solidification rate

Table 4.5   Parameters of casting of circular ingots of aluminium alloys in the ultrasonic field.

Ingot diameter, mm	Casting rate, mm/s	Number of ultrasound sources
65–100	3–4	1
100–200	0.5–3	1
200–300	0.6–1.5	1
300–400	0.4–0.6	1–3
400–500	0.3–0.4	3–5
600–1200	0.2–0.3	7–9

Figure 4.21   One of the largest ingots (1200 mm diameter, 10 ton weight) of alloy 1161 (2324) with non-dendritic structure obtained by casting with UST.

and the mechanical properties are inferior to the properties of continuous casting ingots. At the same time, many important shaped workpieces, such as turbine discs with thin vanes, etc., can be obtained only by shaped casting technology.

The reduction of hydrogen content in an alloy and the increase in density of cast metal are the important factors for raising the quality of shaped castings. The use of UST in this case enables one to increase effectively the density of castings made of aluminium alloys and, as a result of the homogenizing influence on the melt, to improve the chemical homogeneity of castings.

As we have already pointed out, the ultrasonic degassing of liquid aluminium alloys in the mixer was the first industrial application of ultrasonic cavitation treatment [2]. For this purpose special ultrasonic installations – ultrasonic degasators – have been developed to treat melts of 150–300 kg weight with ultrasound in crucible mixers.

Studies and industrial practice have shown that the cavitation treatment of a melt before solidification of the liquid metal in a casting mould markedly changes the main physicochemical properties of shaped castings obtained from a great diversity of cast aluminium alloys. In this case, in addition to the twofold reduction in the hydrogen content (Table 4.6) and the significant increase in the density of the metal (from rank 4 to ranks 1–2 in the porosity scale), the homogenizing effect of cavitation treatment on the melt seems to take on a great importance. Upon ultrasonic degassing, the characteristics of strength and plasticity rise simultaneously. This is particularly clear from comparison of the efficiency of the most widely used industrial degassing methods employed in shaped-casting production. In Table 4.6 the dependence of the main physicochemical characteristics on degassing technique is illustrated by castings made of alloy AL4 (A361).

Table 4.6 Comparison of industrial degassing procedures of alloy AL4 (A361).

Degassing procedure	Hydrogen content, cm^3/100 g	Density, g/cm^3	Porosity	Mechanical properties	
				σ_u, MPa	δ, %
Ultrasonic degassing	0.17	2.706	1–2	245	5.1
Vacuum treatment	0.20	2.681	1–2	228	4.2
Argon blow	0.26	2.667	2–3	233	4.0
Hexachloroethane refining	0.30	2.665	2–3	212	4.5
Universal-flux refining	0.26	2.663	3–4	225	4.0
Initial (non-refined) melt	0.35	2.660	4	200	3.8

As follows from Table 4.6, ultrasound degassing significantly increases the density of cast metal and enables one to obtain almost poreless castings (rank 1 in the porosity scale). The experience of several large plants, which produce intricate shaped castings from aluminium melts and where ultrasound degassing is implemented, shows that this technology almost entirely eliminates porosity in castings, which is usually detected at the final stages of mechanical processing of shaped castings.

Table 4.7 Flowability of melt in a spiral as a function of the degassing procedure for some aluminium cast alloys.

Degassing procedure	Spiral length, mm		
	Cast aluminium alloy		
	AL9 (A356)	AL3 (A356 type)	ATsR (Al–Ce)*
Initial melt (without degassing)	500	500	600
Argon blow	550	600	600
Ultrasonic degassing	670	670	720

*Eutectic alloy

It was also found that degassing cavitation treatment of melts appreciably improves the structure and mechanical properties of castings. This is best seen in castings of an intricate shape obtained by precise casting procedures in ceramic and gypsum moulds when alloys solidify with a rather low rate and the structure is usually coarsened. The homogenizing (refining) effect of ultrasonic treatment on the melt favours the improvement of the structure and properties of castings uniformly over the entire cross-section.

The positive effect of UST on the shaped casting process is also revealed in the improvement of filling intricate shaped moulds in passing from large to small cross-sections. As the main indicator we can use the enhancement of capability of molten metal to flow, which is caused by refining the metal not only from hydrogen but also from solid oxide inclusions. In Table 4.7 we present data on the influence of various degassing procedures on the flowability of melt for several alloys. A comparison was made according to the regular technique, by measuring the length of a spiral filled with metal.

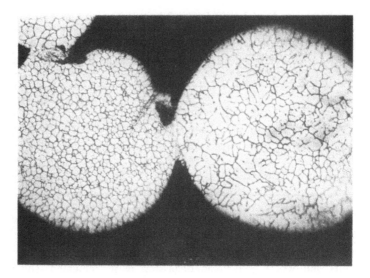

Figure 4.22   Microstructure (×50) of two granules – with non-dendritic (at the left) and dendritic (at the right) grains – adhered in the process of their atomization.

4.5.3. Rapid solidification of aluminium alloys

As the cooling rate, used for melt solidification, rises up to 1000°C/s and to higher values, as compared to the cooling rates < 100°C/s realized in continuous casting and in casting into metallic moulds, it becomes possible to significantly improve the alloy properties due to the solution of excess amounts of transition metals in the solid solution. Small metal particles – granules of 50–1000 μm size – obtained in this case by various methods are then briquetted and subjected to hot deformation for manufacturing semi-finished products of different kinds. This technology, called in Russia granular metallurgy, provides a qualitatively new level of properties of aluminium alloys [21]. Methods for forming granules of aluminium alloys are very diverse – from centrifugal spraying in an inert atmosphere to gas-jet spraying of the melt. Sometimes acoustic jet nozzles are used. Application of ultrasound oscillations to such a nozzle results in the periodic supply of a gas jet to the melt, which provides additional refinement of the granules and enhances the homogeneity of their dimensions. As a result, the structure of extruded articles becomes more fine-grained and their properties are improved.

To obtain anomalously oversaturated solid solutions of transition metals in granulated aluminium alloys it is necessary to have a homogeneous initial melt, which, according to the data presented in Section 2.1, requires heating to 1000–1200°C and higher. Thus, granular metallurgy of aluminium alloys calls for melt *homogenization* as one of the obligatory conditions.

At higher cooling rates, along with the mechanism of heterogeneous solidification, melt supercooling ahead of the crystallization front and homogeneous nucleation begin to play a significant part. In this case, the best grain refinement of the structure and the formation of non-dendritic grains may often be observed close to the solidification of granules with dendritic structure (Fig. 4.22).

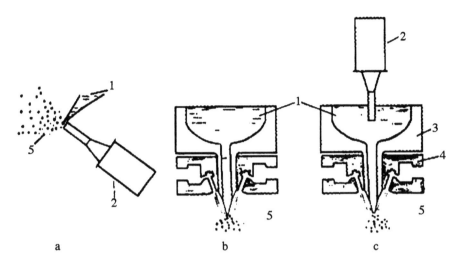

Figure 4.23 Schematic views of rapid solidification of aluminum alloys: (a) ultrasonic atomization in a thin layer; (b) atomization with the use of an acoustic (ultrasonic) atomizer; (c) combined atomization with the use of an acoustic atomizer and preliminary cavitation treatment of the atomized melt. (1) melt, (2) ultrasound source, (3) funnel, (4) acoustic atomizer, (5) granules.

Ultrasonic treatment of melts with development of cavitation processes conducted before the onset (or in the process) of rapid solidification allows the non-dendritic structure to be obtained more reliably in almost all granules.

Figure 4.23 shows the principal schemes of UST of melts in the granulation procedure: spraying in a thin layer at the surface of the ultrasound source, spraying with the use of an ultrasound gas nozzle, and the double UST of melts (gas spraying with ultrasound frequency after preliminary cavitation treatment).

We set forth here our experimental results [4] concerning the rapid solidification of 1960 (7055) alloys of the Al–Zn–Mg–Cu–Zr system with enhanced Zr content, as well as alloys of the supral type belonging to the Al–Cu–Zr system. Granules of 50–150 μm size were obtained by the method of ultrasonic spraying in a thin layer (Fig. 4.23a). At moderate Zr concentrations, a sufficiently high cooling rate, and an amplitude of the ultrasound instrument providing developed cavitation in the thin layer, granules formed in cooling have, as a rule, a non-dendritic structure. An increase in Zr concentration in alloys (at constant cooling rate and constant parameters of ultrasonic atomization) favours the increase of the proportion of granules with non-dendritic grains. It was also established that the proportion of granules with non-dendritic grain in small fractions of granules (100 μm and less) is greater than in large fractions.

This is consistent with the known fact that an increase in the cooling rate causes an increase in the shift of the peritectic point on the phase diagram. For example, for the Al–Zn–Cu–Mg–Zr system the saturation point is shifted to 1% Zr, while for the Al–Cu–Zr system it is displaced to 1.3% Zr. When these zirconium concentrations are achieved, we observe a dramatic growth of the concentration of primary intermetallics which serve as nuclei of solidification of the solid solution. With an increase in the cooling rate, i.e. with a decrease in the droplet size, melt overcooling before solidification grows and progressively smaller intermetallic particles become the effective centres of phase transformation. As a result, an increase in the growth rate leads to the refinement of the non-dendritic grain formed inside each granule. Consequently, the general relation between the size of non-dendritic grains and the melt cooling rate in solidification is verified [16].

The development of cavitation processes in ultrasonic spraying favours the activation of crystallization centres. In spite of the high rate of the spraying process, activation of uncontrolled impurities and homogenization of the melt in the ultrasonic field have enough time to occur during 2 or 3 oscillation periods. If the ultrasonic spraying is implemented in nitrogen or helium vapours, the efficiency of formation of non-dendritic grains in granules is additionally enhanced due to the increase in the cooling rate.

The effect of the non-dendritic structure of granules on the properties of extruded and rolled semi-finished products produced from the alloy of the Al–Cu–Zr system was studied. For this purpose batches of granules with dendritic and non-dendritic structures obtained by ultrasonic spraying at various zirconium concentrations in the alloy were selected. Large (1–2 mm) granules obtained by centrifugal granulation in water were taken for comparison. Granules were charged in aluminium cups of 95 mm diameter and 320 mm in height, briquetted at 420°C under a pressure of 760 MPa, and extruded in strips of 5×60 mm cross-section (Table 4.8).

The mechanical properties of extruded strips after heat treatment (quenching in water from 550°C and ageing at 170°C for 6 h) in the longitudinal and transverse directions are presented in Table 4.9.

The data obtained suggest a noticeable influence of the structure of granules on the property of deformed metal. Granules with non-dendritic structure possess the best combination of properties. Sheets of 1 mm thick, obtained by rolling from the same strips, were used to study the superplasticity effect. The results of tests and sizes of recrystallized grains after annealing at 530°C for 0.5 h are shown in Table 4.10.

Thus, ultrasonic cavitation treatment makes it possible to reliably form a non-dendritic structure in granules and has a positive effect on the properties of deformed semi-finished products prepared from these granules.

4.6   Effect of Ultrasonic Treatment of Melts on the Structure and Properties of Deformed Semi-finished Products Obtained from Aluminium Alloys

The hereditary influence of the structure and properties of ingots on the structure and final properties of deformed semi-finished products are described in detail in the literature. It is thought that after 80–85% deformation we can obtain sufficiently high properties of deformed metal, independently of the initial ingot structure. Investigations show, however, that this assertion refers to the deformation of ingots with a fairly rough structure. Refinement of all elements of the cast structure, from the grain size to the sizes of separate phases, the elimination of zonal segregation, the increase in density, and the decrease in the content of hydrogen and solid inclusions have an extremely large effect on the structure and properties of deformed metal, in spite of a high degree of deformation. *This is particularly true for the case of the limiting cast grain refinement and the formation of non-dendritic ingot structure.*

Table 4.8 Characteristics of granules for extrusion.

Batch no.	Zr concentration, %	Production technique	Structure type
1	0.4	Ultrasonic spraying	dendritic
2	1.8	Ultrasonic spraying	non-dendritic
3	1.5	Centrifugal spraying	dendritic

Table 4.9 Mechanical properties of extruded strips made of an alloy of the Al–Cu–Zr system.

Batch no.	Longitudinal direction		Transverse direction	
	σ_u, MPa	δ,%	σ_u, MPa	δ,%
1	363	16.9	460	16.7
2	485	15.6	461	15.0
3	444	11.4	336	4.6

Table 4.10 Effect of the structure of initial granules and the size of recrystallized grains in sheets of an alloy of the Al–Cu–Zr system on their superplasticity.

Batch no.*	Size of recrystallized grains, μm		Elongation, %
	Longitudinal direction	Height direction	
1	500	32	7
2	10	8	164–200
3	100	25	50

*Batches are numbered as in Tables 4.8 and 4.9

In this section we will consider the effect of ultrasonic cavitation treatment of a melt in the refinement processes (degassing and filtering) and the effect of non-dendritic solidification on the structure and properties of deformed semi-finished products and critical welded joints made of these products.

4.6.1. Effect of ultrasonic refining treatment of melts on the structure and properties of deformed semi-finished products and welded joints

The efficiency of ultrasonic refining of melts can be illustrated most distinctly by a welded AMG6 (Al–6% Mg–0.6% Mn) alloy which finds wide technological use. In view of the high magnesium content, the hydrogen concentration in continuous casting of ingots of this alloy is typically high enough to produce segregations after deformation. These segregations facilitate further the formation of defects in welding, thereby reducing the air tightness of critically important articles (vessels for storage of various gases, etc.).

Ultrasonic degassing of the melt in the flow during continuous casting of large-scale flat ingots of AMG6 alloy of 300×1700 mm in cross section enables one to improve the quality of ingots and sheets produced from these ingots (Table 4.11).

Table 4.11 Effect of ultrasonic degassing of the AMG6 alloy melt on the tendency of 10-mm thick sheets to lamination and on fatigue endurance.

Metal characteristics	Degassing procedure	
	Without ultrasound	Ultrasonic degassing
Hydrogen concentration, cm^3/100 g		
in melt	0.60	0.3
in ingot	0.33–0.37	0.20–0.25
in sheet	0.30–0.34	0.18–0.22
Tendency to lamination (hot probe number)	II–III	1
Endurance in fatigue tests before failure, 10^6 cycles (stress is 160 MPa)	0.07–2.61	0.33–0.37

As follows from the data presented above, cavitation degassing treatment of melts significantly raises the quality of rolled products. Equally good or better results were obtained in producing extruded semi-finished products and, in particular, welded wires made of this alloy and used in argon-arc welding.

Figure 4.24 shows the macrostructure of 10-mm thick sheets and welded joints after hot probe tests for presence of laminations. In the course of these tests a specially selected macrotemplate was rapidly heated to a temperature just above the solidus temperature (585°C for AMG6 alloy) and was etched after cooling in 10% alkali solution until the appearance of segregations. It is clearly seen from Fig. 4.24 that the decrease in the hydrogen content in the sheet from 0.33 to 0.21 cm^3/100 g caused by ultrasonic treatment of melt virtually rules out the appearance of segregations. As a consequence, such a metal can be employed for producing hermetic welded tanks without the risk of leaks occurring.

In Table 4.12 we present the mechanical properties of welded joints of 10-mm sheets made of AMG6 alloy obtained by various refinement (degassing) procedures. Argon lancing lowers the hydrogen content to 0.4–0.5 cm^3/100 g, whereas the content of hydrogen after vacuum treatment in the mixer and after ultrasonic treatment of the melt flow en route to the mould is approximately the same – 0.3 cm^3/100 g.

Table 4.12 Effect of the mode of refining (degassing) AMG6 alloy melt on the mechanical properties of welded joints of 10-mm sheets.

Characteristic	Refining mode		
	Argon lancing	Vacuum treatment	Ultrasonic treatment
σ_u, MPa	300	321	325
Sag angle, deg.	57	65	87
Sag, mm	0.28	0.32	0.51

It is clear that the strength and plasticity of welded joints increase after ultrasonic treatment of melts. This effect is particularly pronounced in comparison with the effect achievable in using the method of vacuum treatment of melt prior to its casting in the mould, which is quite popular in domestic and foreign practice. We relate a considerable

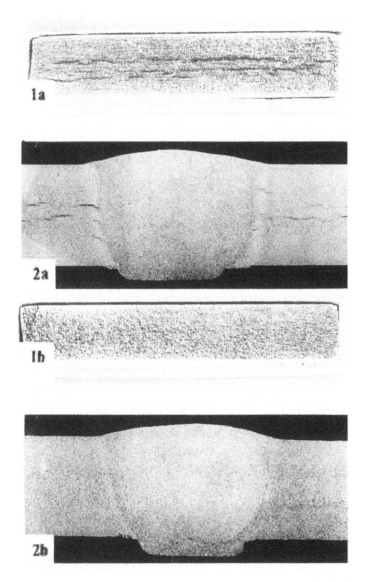

Figure 4.24   Macrostructure (×1.0) of sheets (1) and welded joints (2) of alloy AMG6 after hot probe test: ingot casting without ultrasonic degassing of melt (a) and after ultrasonic degassing (b).

improvement of mechanical properties after UST not only to the refinement of metal from gaseous and solid non-metallic inclusions but also to the *increase in melt homogeneity in the cavitation field.*

The results of tests for air tightness of welded joints of 10-mm sheets of AMG6 alloy are presented in Fig. 4.25. For this purpose, welded joints were tested after argon-arc welding in a special water-filled transparent tank. The presence of a leak was detected by the emergence of bubbles after pressurization of closed vessels with air or argon. The welded

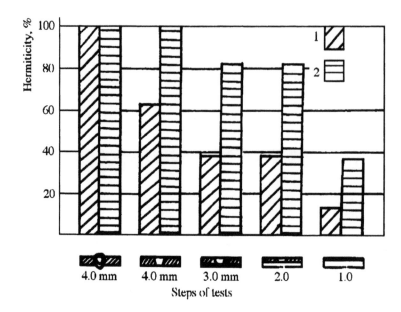

Figure 4.25   Comparative airtightness tests (under pressure in a transparent vessel) of welded joints of alloy AMG6 sheets after sequential turning from 4 to 1 mm without ultrasonic degassing of melt (1) and after ultrasonic degassing (2).

joint thickness was successively reduced from 4 to 1 mm by mechanical treatment of this joint. Clearly, the effect of ultrasonic degassing begins to manifest itself after the first mechanical treatment of the 4-mm thick joint has been made with removal of surplus metal after welding. A particularly significant improvement in air tightness was reached in the case of simultaneous application in the welding process of sheets and welded (added) wire obtained with the use of ultrasonic melt degassing.

4.6.2. Effect of non-dendritic melt solidification on the structure and properties of deformed semi-finished products made of structural aluminium alloys

The effect of the ingot structure on the properties of deformed metal has been studied by many investigators. We can refer, as an example, to the detailed review in the monograph written by Bondarev, Napalkov, and Tararyshkin [22], which is devoted to the problem of aluminium alloy modification. However, in this work, as in many others, the authors used ingots with refined dendritic structure.

In this section we will consider the effect of the non-dendritic ingot structure, i.e. the limiting degree of grain refinement of cast metal, on the structure and properties of ingots.

Small-scale stampings

Small-sized stampings of one of the high-strength 1960 (7055) grade aluminium alloys were produced from ingots of 65–70 mm diameter with dendritic (with 400–500 μm grain

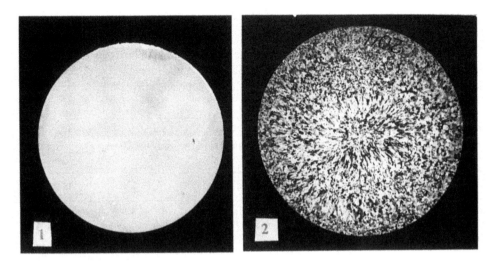

Figure 4.26 Macrostructure (×1.0) of 65–70 mm diameter ingots of alloy 1960 (7055); (1) ingot with non-dendritic structure after UST, (2) ingot with dendritic structure (without UST).

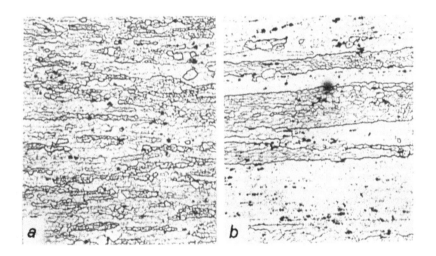

Figure 4.27 Microstructure (×250) of the central part of small stampings of alloy 1960 obtained respectively from 65–70 diameter ingots with UST (a) and without UST (b); see Fig. 4.26.

size) and non-dendritic (with 20–30 μm grain size) structures. The mechanical properties of both types of stampings were examined after strengthening heat treatment (quenching followed by ageing).

Substantial differences in the macrostructure of the initial ingots and stampings produced from these ingots in the most dangerous height direction are illustrated by Figs. 4.26 and 4.27. The mechanical properties of the indicated stampings in this direction are presented

234   I. G. BRODOVA *et al.*

in Table 4.13. The results of tests enable one to assess the effect of the non-dendritic ingot structure on the structure of stampings (Fig. 4.27) and their mechanical properties in the most critical direction. It is quite obvious that the use of ingots with non-dendritic structure enhances simultaneously the strength and plasticity of stamped metal: at sufficiently high strength (more than 600 MPa) plasticity is almost doubled, which means the rise of structural strength or "viability" of this article.

Table 4.13 Effect of the non-dendritic ingot structure on the properties of stampings of 1960 alloy in the height direction (the average properties determined from the results of 10 tests are indicated in parentheses).

Characteristic	Non-dendritic structure	Dendritic structure
σ_u, MPa	646–680 (668)	625–649 (630)
δ, %	4.9–5.9 (5.3)	2.1–4.3 (3.3)

Large-scale stampings

The demand for large-scale stampings (of up to 5 m in length) made of high-strength aluminium alloys arose in connection with designing and producing broad-fuselage airplanes. This, in turn, called for the development of technology for casting large-sized ingots of 845–960 mm in diameter produced from 7050- and 2324-type alloys. As we have already pointed out, the use of UST of melts in the solidification process made it possible to obtain such ingots without cracks and to form in them a non-dendritic structure.

Figure 4.28 shows the macrostructure of one piece of this intricate-shaped stamping obtained from an ingot with non-dendritic structure. We see that the limiting cast grain refinement permitted high homogeneity of the refined structure to be retained throughout the cross-section even after deformation.

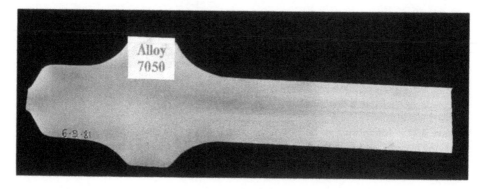

Figure 4.28 Macrostructure (×0.5) of a fragment of large stamping of alloy 7050 obtained from an 830 mm diameter ingot with non-dendritic structure.

The mechanical properties of stampings of this type with dendritic and non-dendritic structures are presented in Table 4.14.

Table 4.14 Effect of the non-dendritic ingot structure on mechanical properties of large-scale stampings made of a high-strength 7050-type alloy after strengthening thermal treatment (all results are given after averaging over 2–4 sections of castings, with 10 tests in each section; L is the longitudinal direction, T is the transverse direction, and H is the height direction).

Characteristic	Non-dendritic structure			Dendritic structure		
	L	T	H	L	T	H
σ_u, MPa	476	476	463	499	488	461
$\sigma_{0.2}$, MPa	410	406	400	427	414	405
δ, %	17.0	15.2	8.8	10.5	11.7	5.6
φ, %	52.2	36.6	30.0	41.0	25.0	10.1
K_{1C}, MPa m$^{1/2}$	183	–	103	138	–	96

Large-scale extruded panels

The production of broad-fuselage airplanes required extruded panels of up to 30 m length. They were obtained from large, round ingots of up to 960 mm in diameter. Figure 4.29 shows the macrostructure of fragments of such panels obtained from ingots with dendritic

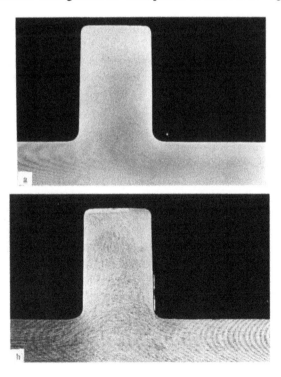

Figure 4.29 Macrostructure (×0.5) of fragments of large long extruded panel of alloy 1973 (7050) obtained from 830 mm diameter ingots with non-dendritic (a) and dendritic (b) structures.

and non-dendritic structures and Table 4.15 contains the data of tests of large-scale extruded panels obtained with the use of vacuum treatment in a mixer and additional ultrasonic treatment for obtaining a non-dendritic ingot structure.

Table 4.15 Effect of the non-dendritic structure on the quality of long panels of 1973 (7050) alloys (averaged test data are indicated in parentheses).

Technology	Mechanical properties in height direction		Number of defects revealed by ultrasonic control
	σ_u, MPa	δ, %	
Vacuum treatment	465–489 (463)	0.8–3.5 (2.6)	more than 150
Additional ultrasonic treatment	440–478 (478)	3.3–4.0 (3.5)	no more than 20

It is easily inferred from the data in Table 4.15 that the non-dendritic structure has a quite positive effect on the properties of extruded panels.

The examples presented here seem to be sufficient to conclude in general about the significant contribution of the limiting refinement of the ingot structure provided by UST of melts on the quality of deformed metal, independently of the conditions and degree of deformation.

4.7 Ultrasonic Equipment for Liquid Aluminium Alloy Processing

More than 40-years of experience of using various types of ultrasonic equipment in experimental and industrial conditions allows us to recommend a series of the most effective systems for use in metallurgical practice.

4.7.1. Industrial ultrasound sources

Industrial ultrasonic thyristor generators (UZG2-4M, UZG3-4, and others) are designed to power a rod-type magnetostrictive transducer (PMS-15A-18) in the acoustic feedback mode, i.e. in the automatic regime of ultrasound emission into the melt. A more powerful thyristor generator USG3-10 can be used to power two such transducers simultaneously. In recent years, thyristor generators have completely replaced their tube analogs which were widely used in the past. Their instant readiness to work (without preliminary tube heating), air cooling, simplified maintenance, and other positive features of ultrasonic thyristor generators have helped them to become widely accepted among process engineers.

A typical ultrasonic thyristor generator (UZG3-4) is housed in a metal cabinet (Fig. 4.30) whose front panel contains control and signaling instruments, knobs and switches. The generator consists of a power transformer, inverter, control unit, power supply, protection relays, magnetization current rectifier, and control panel. For illustration we present the technical characteristics of an industrial ultrasonic generator UZG3-4:

- Output power 4.5 kW ± 20%
- Output voltage 360 V ± 20%
- Output frequency (variable) 18 kHz ± 7.5%

Figure 4.30   Commercial ultrasonic thyristor generator UZG-3-4.

- Efficiency 75%
- Power supply voltage (three-phase, with earthed connection) 380 V ± 5
- Mains frequency 50 Hz
- Overall dimensions 660 × 590 × 1425 mm
- Weight 230 kg

The generator design can be remotely controlled from a panel mounted at the casting operator workplace 50 m away from the generator. This allows the generator to be placed conveniently far away from the place of metal casting and its solidification.

A magnetostrictive transducer PMS-15A-18 (Fig. 4.31) is designed to operate with the UZG3-4 generator. Its power requirement is 3 kW at a frequency of 18 kHz.

The performance characteristics of a PMS-15A-18 rod transducer are as follows:
- Power of losses 3.0 kW
- Power supply 360 V ± 80
- Resonant frequency 18 kHz ± 0.5
- Amplitude of waveguide displacement 15 μm
- Overall dimensions:
    - diameter 175 mm
    - height 353 mm
    - weight 13 kg

238   I. G. BRODOVA *et al.*

Figure 4.31   Commercial magnetostrictive transducer PMS-15A-18.

The transducer is readily built in any equipment for ultrasonic treatment of melts.

In order to provide for contact UST of melts, the waveguiding–radiating system is connected to the acoustic transformer. Typically, such a system consists of two main components: the waveguide (oscillation raiser) and radiator (tool) which are connected to each other and to the acoustic transformer. As a rule, the length of each part of this system is equal to or is a multiple of the half-wavelength of sound in the element material. This enables the connection to be made at the point corresponding to the antinode of oscillations.

It is customary to make the waveguide of carbon steel of grade 45 (1045). With the aim of long-term operation in aluminium-based melts, the instrument is produced from a cavitation-resistant high-melting alloy on the base of transition metals (Ti, Nb, etc.). Depending on the kind of technological process, the waveguiding system is composed from various elements of cylindrical, conical, exponential, or step-like shape. In this case it is important to ensure the resonance of the system and transfer into the melt of the maximum acoustic energy required for implementing cavitation processes.

Ultrasonic equipment for melt refinement

The first Russian industrial ultrasonic degassing installation UZD-100 (200) was developed in 1959 to treat 100–200 kg of melt in a crucible furnace before casting in moulds (Fig. 4.32). The degassing unit was developed for overhead monorail travel above several furnaces to degas the melt in each successively. The UZD-200 unit operated in

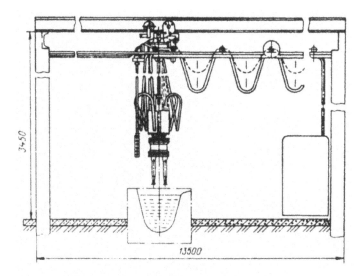

Figure 4.32   (a) Ultrasonic overhead degassing system. (b) Shop view of ultrasonic degassing UZD-100 (UZD-200) unit.

tandem with a 10 kW lamp generator and could be used at a distance of up to 30 m from the latter. Four 2.5 kW magnetostrictive transducers with a frequency of 19.5 kHz were used in the degassing unit and a special switching circuit allowed for their alternate operation 15–20 s apart, thus providing uniform treatment of the entire bulk of the melt. Later, the degassing unit design was modified to operate with modern types of generators and transducers (Figs. 4.30, 4.31).

A powerful installation with 9–12 PMS 15A-18 transducers operating, accordingly, with 9–12 thyristor generators UZG2-4M of 4.5 kW each was developed for ultrasonic degassing of the melt flow during continuous casting of large-scale ingots with a flow rate of 200 kg/min (Fig. 4.33). In contrast to the UZD-200 degassing unit, all 9–12 ultrasound

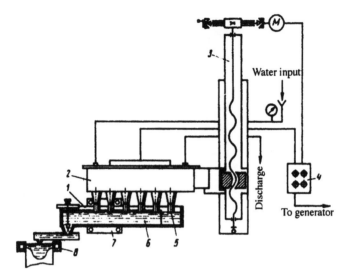

Figure 4.33 Pilot industrial system of ultrasonic degassing of flow (with an efficiency of up to 150 kg/min) incoming from a 40 ton mixer used in casting of large flat ingots: (1) outlet box, (2) arm with ultrasonic sources, (3) arm driving mechanism, (4) terminal box, (5) ultrasonic radiators, (6) melt, (7) notch, (8) mould.

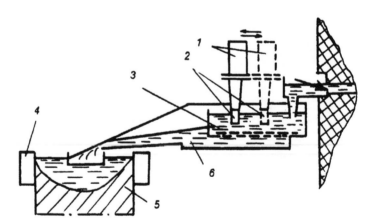

Figure 4.34 One of the schemes of dynamical filtration of melts with multilayered screen filter in an ultrasonic field: (1) ultrasonic source, (2) radiator, (3) multilayered filter, (4) table of casting machine with mould, (5) ingot, (6) melt flow.

sources operated in the course of continuous casting of ingots in the continuous radiation mode, thereby providing effective degassing of the entire flow of the melt.

To combine the degassing process in the cavitation field with fine filtering, fine-filtering installations for implementation of the Uzfirals process were developed. With this purpose, a multilayer glass-cloth filter was placed en route of the melt flow in continuous casting (Fig. 4.34), while an ultrasound source (or several sources) traveled over the surface in such

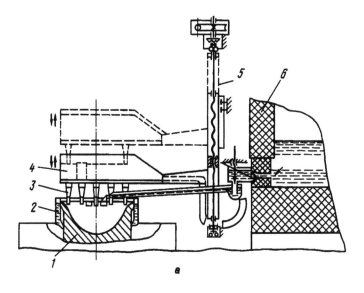

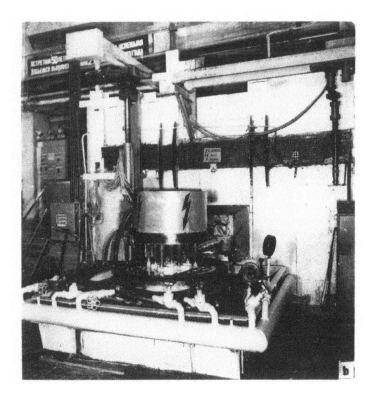

Figure 4.35 Schematic view (a) and shop view (b) of a commercial setup for continuous casting of large ingots of aluminum alloys with UST of melt in a liquid pool of ingot. (1) ingot, (2) mould, (3) ultrasonic radiators, (4) transducer unit, (5) mechanical drive, (6) mixer with melt.

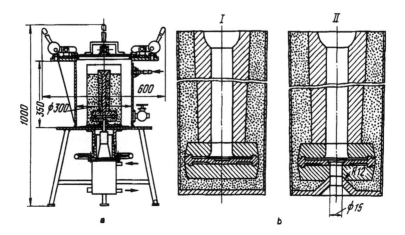

Figure 4.36 (a) Schematic view of precise pressurized casting (in autoclaves) in gypsum–asbestos moulds with ultrasonic treatment. (b) Section views of a gypsum– asbestos mould for casting a turbine wheel in the flask: (I) casting without UST, (II) casting with UST.

Figure 4.37 Mould precise castings of aluminum alloys obtained with UST in ceramic moulds (casting in autoclave).

a way as to activate the maximum possible area. Depending on the particular casting conditions, the source makes a reciprocal or circular motion over the filter.

The equipment for ultrasonic treatment of melt in continuous or shaped casting processes

To influence the process of ingot solidification or shaped casting it is desirable for the cavitation zone to be approached as near as possible to the crystallization front. To do this, ultrasonic treatment is accomplished, as a rule, directly in the liquid pool of an ingot during continuous casting. A few ultrasound sources are used for uniform treatment of the entire section of round or flat large-scale ingots (see Table 4.5). The principal scheme (a) and appearance (b) of an industrial ultrasonic installation for continuous casting of large-scale ingots of 800–1200 diameter with non-dendritic structure are shown in Fig. 4.35. The appearance of one of the largest ingots of 1200 mm diameter and 10 ton weight was shown earlier in Fig. 4.21.

In multimould casting of small and medium ingots, it is appropriate to perform ultrasonic treatment in a special hot top mould, using, for example, Air-slip technology. In some cases it is possible to combine the technology of casting into an electromagnetic mould with ultrasonic treatment of the melt in a liquid pool of the ingot [4].

Special autoclaves with introduction of ultrasound oscillations in the casting mould from the bottom upwards can be employed for precise casting of shaped castings of critical significance with ultrasonic treatment of melt in the mould (Fig. 4.36a). With this aim, the flask with a gypsum–asbestos mould has a hole for the ultrasonic instrument (on the right in Fig. 4.36b).

Figure 4.37 shows a series of precise castings of critical significance produced from aluminium alloys with ultrasound action on the solidification process.

******

The data presented in this chapter on the effect of ultrasonic treatment in the developed cavitation mode on the structure and properties of ingots, castings, granules, and deformed semi-finished products give evidence of the high efficiency of this technology.

The diversity of effects of cavitation treatment of melts – the removal of hydrogen and solid non-metallic inclusions, grain refinement up to the grain size corresponding to the formation of a non-dendritic structure, and so on – points to the deep hereditary linkage between the action upon liquid metal and properties of metal. Cavitation treatment of the melt homogenizes almost instantly the liquid metal (over 2–3 periods of sound wave, or in 100–150 μs) improving its structure and properties in cast and deformed conditions.

Ultrasonic treatment of melts is a technological process *alternative* to heat–time treatment (HTT). To some extent, this treatment is superior to HTT in efficiency and usefulness because it can be performed without additional heating of the melt or changing basic technological parameters. At the same time, its implementation requires special ultrasonic equipment, which makes the technological process somewhat more complicated. For this reason, the user chooses, as a rule, the most appropriate technology of treatment of melt. The authors hope that the material of the present chapter contains sufficient information to make such a choice.

Conclusion

This monograph is designed to draw the attention of metal scientists and metallurgists to the problem of controlling the structure and properties of aluminium alloys by means of various external actions upon liquid and solidifying metal.

Such a possibility has opened up in connection with the diversity of structural rearrangements possible in actual microheterogeneous melts under the influence of directed alloying, heat-time and ultrasonic treatment, and application of other means of external action. As was pointed out above, the ideas about the possibility of such rearrangements set forth in Chapter 1 are based in some cases on the results of indirect experiments and are to some extent hypothetical. Nevertheless, the development of experimental technique in conditions of growing interest to the melt structure brings us ever closer to a proper understanding of the real situation in this field.

Whilst we were preparing this book for publication our foreign colleagues (M. Calvo-Dahlborg and U. Dahlborg) studied the structure of the eutectic Sn−Pb melt by small-angle neutron scattering (SANS) method. They confirmed unambiguously the presence in the system of spherical Pb-rich formations. These formations are destroyed irreversibly at temperatures which we supposedly interpreted as the signs of such a rearrangement. Moreover, another form of microheterogeneity of this melt was also revealed − namely, large-scale inclusions of the second phase, which are of a fractal nature. They can be treated as a net of interrelated lead atoms threading the volume enriched with tin. There are grounds to believe that on heating, or under other external influence, this net may decay into separate "macromolecules" similar to protein molecules, which in turn may change their shape, shrink into globules, or dispergate, etc. Thus, the picture of melt microheterogeneity turns out to be more complicated than we imagined earlier. At the same time new data also open up new possibilities for controlling the structure and properties of metals in the liquid state and after solidification.

In Chapters 2 and 3 of this monograph we considered sequentially the effect of heat−time treatment of melt on the structure of cast metal and used some examples to illustrate the influence of such treatment on their properties. We would like to believe that we have convinced the reader of the efficiency of such an influence. However, we should warn specialists not to use directly the data contained in Chapter 2 about the temperatures of homogenization of various aluminium alloys for optimizing technological regimes of their melting in various industrial plants. The point is that insignificant variations in chemical composition of charge materials may considerably affect the conditions of transition of the melt from the microheterogeneous to true solution state. Consequently, it is necessary to determine the homogenization temperatures directly on the specimens produced in the given plant. As we have already pointed out, this requires investigation of the properties of liquid metal combined with obligatory determination of the chemical composition of the specimens before and after the action studied. At present such investigations are conducted only in separate, primarily Russian, laboratories. However, the trend has been toward the expansion of such studies.

Comparison of the wide possibilities of external influence on melts with the results actually obtained in this direction shows that the use of reserves of such actions has just begun.

In particular, no study was made of the effect of structural transformations in pure liquid metals on their structure and properties after solidification. Seemingly, the solution of this problem is of limited significance and is only important to optimize the technology of melting of one-component melts. However, if we also take into consideration the possibility of existence of regions in multicomponent microheterogeneous metallic liquids which are close in composition to pure components, it will be apparent that such transitions can also influence their properties (such as homogenization temperatures).

A broad range of problems related to the optimization of regimes of heat–time (homogenizing) treatment of liquid metals also remains unsolved. In particular, the effect of pre-homogenization structural transitions of melts from one microheterogeneous state to another, also microheterogeneous, on the structure and properties of ingots and castings is poorly known. Recently we have performed a comparative study of amorphous ribbons of Fe–B and Fe–Co–B alloys, which were quenched from temperatures above and below such transitions, and we have found considerable differences in their properties sensitive to microheterogeneity of the amorphous structure (ultimate strength, magnetic characteristics, thermal stability, etc.). It is likely that such differences can also be detected to a greater or lesser degree after solidification of specimens at comparatively low cooling rates.

The study of the influence of physical melt treatment methods (electromagnetic mixing, vibration, and others) is also of great practical and theoretical interest.

In Chapter 4 we considered in detail the effect of ultrasonic treatment of a melt in cavitation mode on the structure and properties of cast and deformed metal. Cavitation treatment of a melt without additional heating provides almost instantaneous homogenization of the liquid metal and creates conditions for considerable change of the structure and properties of cast and deformed articles made of aluminium alloys. A considerable advantage of this technology is the actual possibility of its practical implementation. The material of Chapter 4 gives a number of examples of practical application of ultrasonic treatment in continuous and shaped casting technologies. We think that ultrasonic treatment of melts is an alternative to heat–time treatment, which can fill an important place in the industrial technology of smelting and casting of aluminium alloys.

Naturally, the list of possible methods influencing melts with the aim of raising the quality of cast and deformed metal is not limited by the material of the present monograph.

The authors are sure that in time the homogenizing treatment of liquid metal will become as obligatory a constituent of the technological process as refining the melt before solidification and heat treatment of ingots and castings. We hope that this monograph will draw the reader's attention to this problem and stimulate investigations which will bring us closer to a complete understanding of the processes involved.

References

Introduction

1. Danilov, V.I. (1956) *Stroenie i kristallizatsiya zhidkostei* (Structure and Solidification of Liquids), Kiev: Izd. Akad. Nauk USSR.
2. Frenkel, Ya.I. (1945) *Kineticheskaya teoriya zhidkostei* (Kinetic Theory of Fluid), Moscow: Izd. Akad. Nauk SSSR.
3. *Physics of Simple Liquids* (1968), Ed. Temperley, H.N.V., Rowlinson, J.S., and Rushbrooke, G.S., Amsterdam: North-Holland.
4. Eiring, H., Ree, T., and Hirai, H. (1958) Significant structures in the liquid state. 1, *Proc. Nat. Acad. USA*, **44(7)**, 683.
5. Ubbellohde, A.R. (1965) *Melting and Crystal Structure*, Oxford: Clarendon.
6. Ubellohde, A.R. (1978) *The Molten State of Matter*, London: J. Wiley.
7. Dutchak, Ya.I. (1977) *Rentgenografiya zhidkikh metallov* (X-Ray Diffraction Analysis of Liquid Metals), L'vov: Vyshcha Shkola.
8. Regel, A.P. and Glazov, V.M. (1982) *Zakonomernosti formirovaniya struktury elektronnykh rasplavov* (Regularities in the Structure Formation of Electronic Melts), Moscow: Nauka.
9. Ershov, G.S. and Poznyak, L.A. (1985) *Mikroneodnorodnost' metallov i splavov* (Microheterogeneity of Metals and Alloys), Moscow: Metallurgiya.
10. Bernal, J. (1961) Geometric Approach to the Structure of Liquids, *Usp. Khim.*, **30(10)**, 1312.
11. Waseda, J. (1980) The Structure of Non-Crystalline Materials. Liquid and Amorphous Solids, McGraw-Hill.
12. Polukhin, V.A. and Vatolin, N.A. (1985) *Modelirovanie amorfnykh metallov* (Modelling of Amorphous Metals), Moscow: Nauka.
13. Zhukova, L.A. and Popel, S.I. (1982) Icosahedral model of liquefied inert gases and liquid metals with fcc premelting structure, *Zh. Fiz. Khim.*, **56(2)**, 476.
14. Shechtman, D., Blech, J., Gratias, D., and Cahn, L.W. (1984) Metallic phase with long-range orientational order and no translation symmetry, *Phys. Rev. Lett.*, **53(20)**, 1951.

Chapter 1

1. Skryshevskii, A.F. (1971) *Strukturnyi analiz zhidkostei* (Structural Analysis of Liquids), Moscow: Vyssh. Shkola.

2. Gel'chinskii, B.R., Ancharova, L.P., Ancharov, A.I., and Shatmanov, T.Sh. (1987) *Nekotorye experimental'nye i chislennye metody issledovaniya struktury blizhnego poryadka* (Some Experimental and Numerical Methods for Studying the Short-range Order Structure), Frunze: Ilim.
3. Borovskii, I.B., Vedrinskii, R.V., Kraizman, V.L., and Sachenko, V.P. (1986) EXAFS spectroscopy as a new method of structural studies, *Usp. Fiz. Nauk*, **149(2)**, 275.
4. Basin, A.S. (1970) Density and thermal expansion of rubidium and cesium in the liquid state up to 1300°C, in: *Issledovaniya teplofizicheskikh svoistv veshchestva* (Study of Thermal Properties of Matter), 81.
5. Popel, P.S., Konovalov, V.A., and Porotov, A.V. (1981) On the accuracy of absolute density measurements by the gamma-ray method, in: *Gamma-ray Method in Metallurgical Experiment*, Novosibirsk: ITF SO Akad. Nauk SSSR, 55.
6. Bibik, E.E. (1981) *Reologiya dispersnykh sistem* (Rheology of Disperse Systems), Leningrad: LGU.
7. Shvidkovskii, E.G. (1955) *Nekotorye voprosy vyazkosti rasplavlennykh metallov* (Some Problems in Viscosity of Molten Metals), Moscow: Gostekhizdat.
8. Regel, A.P. (1956) An electrodeless method for measuring conductivity and the possibility of its application to the problems of physicochemical analysis, *Zh. Neorg. Khim.*, **1(6)**, 1271.
9. Regel, A.P. and Glazov, V.M. (1980) *Fizicheskie svoistva elektronnykh rasplavov* (Physical Properties of Electronic Melts), Moscow: Nauka.
10. Nikolaev, V.O., Iolin, E.M., Kozlov, E.N., and Tsirkunova, S.E. (1982) *Izv. Akad. Nauk Latv. SSR, Ser. Fiz.-Tekh. Nauk*, No. 5, 118.
11. Sharykin, Yu.I., Glazkov, V.P., Skovorod'ko, S.N., *et al.* (1979) *Dokl. Akad. Nauk SSSR*, **244(1)**, 78.
12. Astapkovich, A.Yu., Iolin, E.M., Kozlov, E.N., *et al.* (1982) *Dokl. Akad. Nauk SSSR*, **263(1)**, 73.
13. Franze, G., Freyland, W., Glaser, W., *et al.* (1980) *Proc. 4th Int. Conf. (LAM 4), J. de Phys.*, **41**, suppl. 8, 194.
14. Gel'chinskii, B.R. and Vatolin, N.A. (1984) *Dokl. Akad. Nauk SSSR*, **277(5)**, 1109.
15. Ivakhnenko I.S. (1985) Specific features of the structure of metallic melts, *Izv. Vyssh. Uchebn. Zaved., Chern. Met.*, No. 5, 17.
16. Dutchak, Ya.I., (1961) *Fiz. Met. Metalloved.*, **11(2)**, 290.
17. Vertman, A.A. and Samarin, A.M. (1969) *Svoistva rasplavov zheleza* (Properties of Iron Melts), Moscow: Nauka.
18. Novokhatskii, I.A., Kisun'ko, V.Z., and Lad'yanov, V.I. (1985) Peculiarities in the manifestation of different types of structural transformations in metallic melts, *Izv. Vyssh. Uchebn. Zaved., Chern. Met.*, No. 9, 1.
19. Fröberg, M.G. and Cakici, T. (1977) *Arch. Eisenhüttenw.*, **48(3)**, 145.
20. Vatolin, N.A. and Pastukhov, E.A. (1977) *Difraktsionnye issledovaniya stroeniya vysokotemperaturnykh rasplavov* (Diffraction Studies on the Structure of High-Temperature Melts), Moscow: Nauka.
21. Slukhovskii, O.I., Lashko, A.S., and Romanova, A.V. (1975) Structural transformations in liquid iron, *Ukr. Fiz. Zh.*, **90**, 1961.
22. Vatolin, N.A. and Polukhin, V.A. (1985) Structural transformations in metallic melts, *Izv. Vyssh. Uchebn. Zaved., Chern. Met.*, No. 7, 1.
23. Sidorov, V.E., Gushchin, V.S., and Baum, B.A. (1984) Magnetic structure of iron containing oxygen, *Phys. stat. sol.(a)*, **85**, 497.
24. Bazin, Yu.A., Zamyatin, V.M., Nasyirov, Ya.A., and Emel'yanov, A.V. (1985) On structural transformations in liquid aluminium, *Izv. Vyssh. Uchebn. Zaved., Chern.*

Met., No. 5, 28.

25. Vatolin, N.A., Pastukhov, E.A., and Sermyagin, V.N. (1975) *Dokl. Akad. Nauk SSSR*, **222(3)**, 641.

26. Emel'yanov, A.V., Nasyirov, Ya.A., and Bazin, Yu.A. (1983) *Izv. Vyssh. Uchebn. Zaved., Fizika*, No. 12, 125.

27. Gel'chinskii, B.R. (1985) Structural transformations in liquid metals: experimental and theoretical data, *Izv. Vyssh. Uchebn. Zaved., Chern. Met.*, No. 7, 16.

28. Belashchenko, D.K. (1970) *Yavleniya perenosa v zhidkikh metallakh i poluprovodnikakh* (Transport Phenomena in Liquid Metals and Semiconductors), Moscow: Atomizdat.

29. Esche, W. and Peter, O. (1956) Bestimmung der Oberflächenspannung an reinem und legiertem Eisen, *Arch. Eisenhüttenw.*, **27(6)**, 355.

30. Tsarevskii, B.V. and Popel, S.I. (1962) Influence of phosphorus and sulphur on surface properties of iron, *Fiz. Met. Metalloved.*, **13(3)**, 451.

31. Arsent'ev, P.P. and Koledov, L.A. (1976) *Metallicheskie rasplavy i ikh svoistva*, (Metallic Melts and Their Properties), Moscow: Metallurgiya.

32. Romanov, A.A. and Kochegarov, V.G. (1964) Influence of small aluminium additions on viscosity of liquid iron, *Izv. Akad. Nauk SSSR, Metallurgiya i Gornoe Delo*, No. 1, 41.

33. Ubellohde, A. R. (1965) *Melting and Crystal Structure*, Oxford: Clarendon.

34. Arsent'ev, P.P., Filippov, S.I., and Lisitskii, B.S. (1970) Electrical conductivity of iron and iron–carbon melts, *Izv. Vyssh. Uchebn. Zaved., Chern. Met.*, No. 3, 18.

35. Ven Li-shi and Arsent'ev, P.P. (1961) Viscosity of iron melts with oxygen and carbon impurities, *Izv. Vyssh. Uchebn. Zaved., Chern Met.*, No. 7, 5.

36. Romanov, A.A. and Kochegarov, V.G. (1963) Study of viscosity and the structure of iron-carbon melts, *Izv. Vyssh. Uchebn. Zaved., Chern. Met.*, No. 3, 89.

37. Lucas, L.D. (1964) Volume specifique de metaux et alliages liquides a hautes temperatures. II. Elements du groupe VIII et alliages a base de fer, *Mem. scient. Rev. Metallurgie*, **61(2)**, 97.

38. Krieger, W. and Trenkler, H. (1971) Aussagen über den Zustand von Eisen–Kohlenstoff–Schmelzen bei hohen Temperaturen auf Grund des Viskositätsverhaltens, *Arch. Eisenhüttenw.*, **42(10)**, 685.

39. Arsent'ev, P.P. and Filippov, S.I. (1974) Viscosity activity of impurities in pure iron, in: *Problemy stal'nogo slitka* (Problems of Steel Ingot), Moscow: Metallurgiya, p. 249.

40. Krieger, W. and Trenkler, H. (1971) Die Deutung der Schmelzstrukturen von Eisen–Kohlenstoff- und Eisen–Nickel-Legierungen aus dem Viskositätsverhalten, *Arch. Eisenhüttenw.*, **42(3)**, 175.

41. Belashchenko, D.K. (1957) On structural features of liquid alloys of some binary systems, *Dokl. Akad. Nauk SSSR*, **117(1)**, 98.

42. Gel'd, P.V., Baum, B.A., and Petrushevskii, M.S. (1973) *Rasplavy ferrosplavnogo proizvodstva* (Melts in Ferroalloy Production), Moscow: Metallurgiya.

43. Narita, K. and Onoye, T. (1971) Viscosity of liquid iron and steel, in: *Proc. Int. Conf. Sci. and Technol. Iron and Steel*, part 1, Tokyo, p. 400.

44. Kupriyanov, A.A. and Filippov, S.I. (1968) Density and structural transformations in iron and iron–carbon alloys, *Izv. Vyssh. Uchebn. Zaved., Chern. Met.*, No. 9, 10.

45. Popov, D.S., Vishkarev, A.F., Khokhlov, S.F., *et al.* (1969) *Izv. Vyssh. Uchebn. Zaved., Chern. Met.*, No. 7, 120.

46. Bodakin, N.E., Baum, B.A., and Tyagunov, G.V. (1977) Viscosity of liquid alloys of the Fe–Ni system, *Izv. Vyssh. Uchebn. Zaved., Chern. Met.*, No. 5, 18.

47. Bodakin, N.E., Baum, B.A., and Tyagunov, G.V. (1978) Viscosity of melts of the

Fe–Co system, *Izv. Vyssh. Uchebn. Zaved., Chern. Met.*, No. 8, 5.

48. Kosilov, N.S., Baum, B.A., Tyagunov, G.V., and Popel, P.S. (1978) *Izv. Vyssh. Uchebn. Zaved., Chern. Met.*, No. 8, 5.

49. Kosilov, N.S., Popel, P.S., Baum, B.A., *et al.* (1979) Volume characteristics of Fe–Co melts, *Izv. Akad. Nauk SSSR, Metally*, No. 6, 75.

50. Popel, P.S. Tyagunov, G.V., Baum, B.A., *et al.* (1985) Investigation of density of iron–chromium melts by the gamma-ray method, *Zh. Phys. Khim.*, **59(2)**, 399.

51. Borovskii, I.B. and Gurov, K.P. (1959) On the theory of transition-metal solid solutions, *Fiz. Met. Metalloved.*, **4(1)**, 187.

52. Gel'd, P.V., Dovgopol, M.P., Dovgopol, S.P., and Radovskii, I.Z. (1977) Magnetic susceptibility and structure of short-range order of iron–carbon melts, *Dokl. Akad. Nauk SSSR*, **226(4)**, 853.

53. Bazin, Yu.A., Klimenkov, E.A., Baum, B.A., and Mariev, S.A. (1979) Short-range order in liquid iron–carbon alloys, *Ukr. Fiz. Zh.*, **24(7)**, 1052.

54. Shvarev, K.M., Gushchin, V.S., Baum, B.A., and Gel'd, P.V. (1979) Optical constants of iron–carbon alloys in the temperature range 20–1600°C, *Teplofizika Vysokikh Temperatur*, **17(1)**, 66.

55. Tyagunov, G.V., Popel, P.S., Kosilov, N.S., *et al.* (1981) Density of melts of the Fe–C system, *Izv. Akad. Nauk SSSR, Metally*, No. 5, 55.

56. Kudrin, V.A., Elanskii, G.N., and Uchaev, A.N. (1981) Influence of structure and properties of metallic melts on steel quality, *Steel*, No. 9, 21.

57. Ivakhnenko, I.S. and Lyakutkin, A.V. (1981) Structure and properties of iron–carbon melts, in: *Gamma-metod v metallurgicheskom eksperimente* (Gamma-ray Method in Metallurgical Experiment), Novosibirsk: ITF SO Akad. Nauk SSSR, 76.

58. Klimenkov, E.A., Tyagunov, G.V., Baum, B.A., and Epin, V.N. (1983), Influence of carbon and oxygen impurities on resistivity of iron at high temperatures, *Izv. Akad. Nauk SSSR, Metally*, No. 3, 32.

59. Popel, P.S., Arkhangel'skii, E.L., and Makeev, V.V. (1995) Density of iron–boron melts, *Vysokotemp. Rasplavy*, **1(1)**, 85.

60. Popel, P.S., Zamyatin, V.M., Domashnikov, B.P., *et al.* (1983) Influence of scandium on the properties of liquid aluminium, *Izv. Akad. Nauk SSSR, Metally*, No. 3, 38.

61. Papastaikoudis, C. and Papadimitropoulos, D. (1981) Influence of the virtual bound state on the low-field Hall coefficient: II. Dilute alloys of Sc, Ti, V, Co and Ni in Al, *Phys. Rev B: Condens. Matter*, **24(6)**, 3108.

62. Popel, P.S., Domashnikov, V.M., Zamyatin, V.M., *et al.* (1983) Influence of small cadmium additions on the density and surface tension of liquid aluminium and its mechanical properties in the solid state, *Izv. Akad. Nauk SSSR, Metally*, No. 5, 59.

63. Koledov, L.A. and Lyubimov, A.P. (1963) Viscosity of dilute aluminium-based metallic melts, *Izv. Vyssh. Uchebn. Zaved., Chern. Met.*, No. 9, 136.

64. Bokareva, N.M., Gotgil'f, T.L., Eretnov, K.I., *et al.* (1965) Viscosity of tin and its alloys with nickel, *Izv. Vyssh. Uchebn. Zaved., Chern. Met.*, No. 9, 8.

65. Mott, N.F. and Alligaier, R.S. (1967) Localized states in disordered lattices, *Phys. stat. sol.*, **21(1)**, 343.

66. Gurov, K.P. and Borovskii, I.B. (1960) On the theory of dilute solid solutions, *Fiz. Met. Metalloved.*, **10(4)**, 513.

67. Borovskii, I.B. and Gurov, K.P. (1959) Study of electronic spectra of dilute solid solutions, *Fiz. Met. Metalloved.*, **7(2)**, 225.

68. Kaizar, F. and Parette, G. (1980) Magnetic-moment distribution and environmental effects in dilute iron-based alloys with V, Cr and Mn impurities, *Phys. Rev. B: Condens. Matter*, **22(11)**, 5471.

69. Hayashi, E. and Shimizu, M. (1980) Distribution of induced spin density in dilute PdFe alloys, *J. Phys. F: Metal Phys.*, **10(10)**, L275.

70. Ishino, M. and Muto, Y. (1985) Electrical resistivity of binary vanadium alloys, *J. Phys. Soc. Japan*, **54(10)**, 3839.

71. Batchelor, J. (1977) The effect of Brownian motion on the bulk stress in a suspension of spherical particles, *J. Fluid Mechanics*, **83**, part 1, 97.

72. Hushin, Z. (1964) Bounds for viscosity coefficients of liquid mixtures by variational methods, *Proc. I.U.T.A.M. Symposium on Second-Order Effects in Elasticity, Plasticity and Fluid Dynamics*, Haifa, 1962, New- York: Pergamon, p. 246.

73. Friedel, J. (1958) On the electronic structure of transition and heavy metals and their alloys, *J. Phys. Rad.*, **19(6)**, 573.

74. Mrosan, E. and Lehman, G. (1978) Electronic states of 3d-transition metal impurities in aluminium, *Phys. stat. sol. (b)*, **87(1)**, K21.

75. Nieminen, R.M. and Puska, M. (1980) 3d-impurities in Al: density functional results, *J. Phys. F: Metal Phys.*, **10(5)**, L123.

76. Deutz, J., Dederichs, P.H., and Zeller, R. (1981) Local density of states of impurities in Al, *J. Phys. F: Metal Phys.*, **11(8)**, 1787.

77. Gyemant, I. and Vasvari, B. (1983) Cluster phase shifts and virtual bound states for V, Mn and Fe impurities in aluminium, *Phys. stat. sol. (b)*, **115(2)**, K83.

78. Steiner, P., Hochst. H., and Hufner, S. (1977) Electronic states of Mn, Co, Ni and Cu impurities in aluminium, *J. Phys. F: Metal Phys.*, **7(4)**, L105.

79. Shopke, R. and Mrosan, E. (1978) Influence of multiple scattering effects on the residual resistivity of 3d-transition metal impurities in aluminium, *Phys. stat. sol. (b)*, **90(1)**, K95.

80. Lautenschlager, G. and Mrosan, E. (1979) Localization of additional screening charge around 3d-transition metal impurities in Al, *Phys. stat. sol. (b)*, **91(1)**, 109.

81. Podloucky, R., Zeller, R., and Dederichs, P.H. (1980) Electronic structure of magnetic impurities calculated from first principles, *Phys. Rev. B: Condens. Matter*, **22(12)**, 5777.

82. Dagens, L. (1979) Virtual bound states in dilute noble metal based disordered alloys according to the resonant model potential theory, *J. Phys. F.: Metal Phys.*, **9(1)**, 45.

83. Barnard, R.D. and Abdel Rahiem, A.E.E. (1981) Giant low-field magnetoresistance and Hall effect in dilute Al–V and Al–Ti alloys, *Physica B+C*, **107(1–3)**, 505.

84. Barnard, R.D. (1983) The effect of clustering on the Hall and magnetoresistance coefficients in very dilute Al–Ti alloys, *J. Phys. F: Metal Phys.*, **13(3)**, 685.

85. Maurer, M., Cadeville, M.C., and Sanchez, J.P. (1979) Electronic structure of dilute $Fe_{1-x}P_x$ alloys, *J. Phys. F: Metal Phys.*, **9(2)**, 271.

86. Child, H.R. and Cable, J.W. (1976) Temperature dependence of the magnetic-moment distribution around impurities in iron, *Phys. Rev. B: Solid State*, **13(1)**, 277.

87. Rowland, T.J. (1960) Nuclear magnetic resonance in copper alloys. Electron distribution around solute atoms, *Phys. Rev.*, **19(3)**, 900.

88. Kruglov, V.F., Verkhovskii, S.V., and Kleshchev, G.V. (1983) High-temperature studies of nuclear magnetic resonance of ^{27}Al in dilute aluminium–zinc alloys, *Fiz. Met. Metalloved.*, **55(3)**, 617.

89. Brettel, J.M. and Heeger, A.J. (1967) Nuclear magnetic resonance in dilute alloys of Mn, Fe and Cu in aluminium, *Phys. Rev.*, **153(2)**, 319.

90. von Meerwall, E. and Rowland, T.J. (1972) ^{51}V quadrupolar effects in V-transition-metal alloys and solutions of C and N in V, *Phys. Rev. B: Condens. Matter*, **5(7)**, 2480.

91. von Meerwall, E. and Schreiber, D.S. (1971) Local magnetic fields in the

vanadium–manganese alloy system, *Phys. Rev. B: Condens. Matter*, **3(1)**, 1.

92. Dovgopol, S.P. and Zaborovskaya, I.A. (1982) *Elektronnaya struktura, magnetizm i stabil'nost' faz 3d-metallov i splavov v tverdom i zhidkom sostoyaniyakh: Obzory po teplofizicheskim svoistvam veshchestv* (Electronic structure, magnetism and stability of 3d-metal phases and alloys in the solid and liquid states: Reviews on thermal properies of substances), No. 2, Moscow: IVT Akad. Nauk SSSR.

93. Fairlie, R.H. and Greenwood, D.A. (1983) A density–functional approach to the electronic structure of dissolved impurities in simple liquid metals, *J. Phys. F: Metal Phys.*, **13(8)**, 1645.

94. Gel'd, P.V., Singer, V.V., Sandratskii, L.M., *et al.* (1987), Impurity effects in liquid alloys of iron with Mo, W, Nb and Ta, *Dokl. Akad. Nauk SSSR*, **292(3)**, 612.

95. Kurnakov, N.S. (1960) *Selected Works*, vol. 1, Moscow: Izd. Akad. Nauk SSSR.

96. Roll, A. and Motz, H. (1957) Der elektrische Widerstand geschmolzener Kupfer–Zinn-, Silber–Zinn- und Magnesium–Blei-Legierungen, *Z. Metallkd.*, **48(8)**, 435.

97. Roll, A. and Motz, H. (1957) Der elektrische Widerstand von Legierungschmelzen der Mischkristallsysteme Silber–Kupfer, Zinn–Zink und Aluminium, *Z. Metallkd.*, **48(9)**, 495.

98. Okajima, K. and Sakao, H. (1983) Volumes of mixing of liquid alloys, *Trans. Jap. Inst. Metals*, **24(4)**, 223.

99. Bazin, Yu.A., Epin, V.N., Klimenkov, E.A., and Baum, B.A. (1981) Electrical resistivity of liquid alloys of iron with cobalt and nickel, *Izv. Vyssh. Uchebn. Zaved., Fizika*, **24(7)**, 112.

100. Govorukhin, L.V., Klimenkov, E.A., Baum. B.A. *et al.* (1984) Electrical resistivity of iron alloys with chromium and oxygen at high temperatures, *Ukr. Fiz. Zh.*, **29(2)**, 291.

101. Sidorov, V.S., Gushchin, V.S., and Baum, B.A. (1985) Magnetic susceptibility of Fe–Cr–O alloys, *Izv. Vyssh. Uchebn. Zaved., Chern. Met.*, No. 8, 100.

102. Kudryavtseva, E.D., Dovgopol, M.P., Radovskii, I.Z., *et al.* (1980) Influence of composition on electrical resistivity of liquid iron–chromium alloys, *Zh. Fiz. Khim.*, **54(1)**, 145.

103. Komarek, K.L. (1977) Thermodynamik von Legierungsschmelzen: Messmethoden und Ergebnisse, *Ber. Bunsenges. Phys. Chem.*, **81(10)**, 936.

104. Tanji, Y., Mriya, H., and Nakagawa, Y. (1978) Anomalous concentration dependence of thermoelectric power of Fe–Ni (fcc) alloys at high temperatures, *J. Phys. Soc. Jap.*, **45(4)**, 1244.

105. Aldred, A.T., Rainford, B.D., Kouvel, J.S., and Hicks, T.J. (1976) Ferromagnetism in iron–chromium alloys. II. Neutron scattering studies, *Phys. Rev. B: Solid State Phys.*, **14(1)**, 228.

106. Dubiel, S.M. and Zukrowski, J. (1975) Magnetic hyperfine field distributions from Mössbauer spectra of iron–chromium alloys, *Acta phys. pol.*, No. 2, 315.

107. Andronov, V.M., Ivanov, I.G., Sirenko, A.F., and Sherstyuk, V.A. (1987) Oscillatory character of composition–property dependence for metallic solutions, in: *Voprosy atomnoi nauki i tekhniki. Fizika radiatsionnykh povrezhdenii i radiatsionnoe materialovedenie* (Problems of Atomic Science and Technology. Physics of Radiation Damage and Radiation Materials Science), No. 2(40), 83.

108. Kim, D.J. (1970) Ferromagnetism of transition metal alloys, *Phys. Rev. B: Condens. Matter*, **1(9)**, 3725.

109. Hausch, G. and Warlimont, H. (1971) Structural inhomogeneity in Fe– Ni invar alloys studied by electron diffraction, *Phys. Lett. A*, **36(5)**, 415.

110. Wilson, D.R. (1972) *Structure of Liquid Metals and Alloys*, Moscow: Metallurgiya.

111. Regel, A.P. and Glazov, V.M. (1978) *Periodicheskii zakon i fizicheskie svoistva elektronnykh rasplavov* (Periodic Law and Physical Properties of Electronic Melts), Moscow: Nauka.

112. Gel'd, P.V. and Gertman, Yu.M. (1960) Volume effects in mixing of liquid silicon and iron, *Fiz. Met. Metalloved.*, **10(5)**, 793.

113. Danilov, V.I. (1956) *Stroenie i kristallizatsiya zhidkosti* (Structure and Solidification of Liquids), Kiev, Izd. Akad. Nauk USSR.

114. Styles, G.A. (1967) Influence of short-range atomic order on nuclear magnetic resonance in liquid alloys, *Advances Phys.*, **16(62)**, 275.

115. Glazov, V.M. and Chizhevskaya, S.N. (1968) Influence of cooling rate and overheating temperature on the degree of supercooling of germanium melt, *Izv. Akad. Nauk SSSR, Neorg. Materials*, **4(2)**, 171.

116. Glazov, V.M. (1970) Influence of cooling rate and overheating temperature on the degree of supercooling of indium antimonide melt, *Izv. Akad. Nauk SSSR, Neorg. Materials*, **6(10)**, 1775.

117. Makeev, V.V. and Popel, P.S. (1990) Volume characteristics of alloys of the Ni−B system in the range from 1100 to 2170 K, *Zh. Fiz. Khim.*, **64**, 568.

118. Taran, Yu.N. and Mazur, V.I. (1978) *Struktura evtekticheskikh splavov* (Structure of Eutectic Alloys), Moscow: Metallurgiya.

119. Danilov, V.I. and Radchenko, I.V. (1937) X-ray scattering in liquid eutectic alloys, *Zh. Exp. Teor. Fiz.*, **7(9−10)**, 1158.

120. Dutchak, Ya.I. (1977) *Rentgenografiya zhidkikh metallov* (X-Ray Diffraction Analysis of Liquid Metals), L'vov: Vyshcha Shkola.

121. Skryshevskii, A.F. (1956) Structure of liquid eutectic Bi−Pb alloy from X-ray diffraction data, *Dokl. Akad. Nauk USSR*, No. 1, 62.

122. Laschko, A.S. and Romanova, A.V. (1958) Structure of binary metallic liquid alloys, *Ukr. Fiz. Zh.*, **3(3)**, 375.

123. Batalin, G.I. and Kazimirov, V.P. (1971) X-ray diffraction study of the structure of Al−Sn alloys in the liquid state, *Ukr. Fiz. Zh.*, **16(3)**, 378.

124. Bublik, A.I. and Buntar', A.G. (1958) Electron diffraction study of the structure of liquid metalls and alloys, *Kristallografiya*, **3(1)**, 32.

125. Sharrah, P.S., Potz, J.I., and Krush, R.F. (1960) Determination of atomic distribution in liquid lead−bismuth alloys by neutron and X-ray diffraction, *J. Chem. Phys.*, **32(1)**, 241.

126. Ebert, H., Höhler, J., and Steeb, S. (1974) Schallgeschwindigkeitsmesung in Bi−Cu-Schmelzen zur Bestimmung der Kompressibilität und der partiallen Strukturfaktoren, *Z. Naturforsch.*, **29a(12)**, 1890.

127. Lashko, A.S. (1959) On the structure of liquid Au−Sn alloy, *Dokl. Akad. Nauk SSSR*, **125(1)**, 126.

128. Lashko, A.S. and Romanova, A.V. (1961) On the X-ray investigation of the structure of liquid metallic alloys of the systems with eutectic, *Izv. Akad. Nauk SSSR. Metallurgiya i Toplivo*, No. 3, 135.

129. Smallman, R.E. and Frost, B.R.T. (1956) An X-ray investigation of the structure of liquid mercury−thallium alloys, *Acta metallurgica*, **4(6)**, 611.

130. Karlikov, D.N. (1958) X-ray investigation of short-range order in liquid solutions of zinc in mercury, *Ukr. Fiz. Zh.*, **3(3)**, 370.

131. Alekseev, N.V. and Evseev, A.M. (1959) Investigation of the structure of liquid Cd−Sn alloys, *Kristallografiya*, **4(3)**, 348.

132. Dutchak, Ya.I., Mykolaichuk, A.G., and Klym, N.M. (1962) X-ray study of the

structure of some metallic liquids, *Fiz. Met. Metalloved.*, **14(4)**, 548.

133. Klyachko, Yu.A. and Kunin, L.L. (1949) On the surface tension of eutectic alloys, *Dokl. Akad. Nauk SSSR*, **64(1)**, 85.

134. Rohl, A. and Uhl, E. (1959) Der elektrische Widerstand geschmolzener Gold–Zinn-, Gold–Blei- und Silber–Blei-Legierungen, *Z. Metallkd.*, **50(3)**, 159.

135. Roll, A. and Fees, G. (1960) Der elektrische Widerstand geschmolzener Blei–Zinn- und Quecksilber–Thallium-Legierungen, *Z. Metallkd.*, **51(9)**, 540.

136. Roll, A. and Swamy, N.K.A. (1961) Der elektrische Widerstand geschmolzener Zweistofflegierungen des Kadmiums mit Blei, Quecksilber und des Antimons mit Wismut, *Z. Metallkd.*, **52(2)**, 111.

137. Roll, A. and Basu, P. (1963) Der elektrische Widerstand geschmolzener Zinn–Antimon- und Zinn–Wismut-Legierungen, *Z. Metallkd.*, **54(9)**, 511.

138. Roll, A. and Bismas, T.K. (1964) Der elektrische Widerstand geschmolzener Blei–Antimon- und Blei–Wismut-legierungen, *Z. Metallkd.*, **55(12)**, 794.

139. Geguzin, Ya.E. and Pines, B.Ya. (1951) Energy of mixing of binary metallic alloys. I. Lead–tin system, *Zh. Fiz. Khim.*, **25(10)**, 1228.

140. Geguzin, Ya.E. and Pines, B.Ya. (1952) Energy of mixing of binary metallic alloys. IV. Bismuth–tin and bismuth–lead systems, *Zh. Fiz. Khim.*, **26(2)**, 165.

141. Evseev, A.M. and Voronin, G.F. (1966) *Termodinamika i struktura zhidkikh metallicheskikh splavov* (Thermodynamics and Structure of Liquid Metallic Alloys), Moscow: Izd. MGU.

142. Ershov, G.S. and Bychkov, Yu.B. (1979) *Vysokoprochnye alyuminievye splavy na osnove vtorichnogo syr'ya* (High-strength aluminium alloys based on secondary raw materials), Moscow: Metallurgiya.

143. Gaibullaev, F. and Regel, A.P. (1957) Peculiarities in temperature dependence of resistivity of liquid eutectic systems, *Zh. Tekh. Fiz.*, **27(9)**, 1996.

144. Lozovskii, V.N., Politova, N.F., and Shutova, E.V. (1968) On the temperature dependence of the coefficient of diffusion of silicon in liquid aluminium, *Fiz. Met. Metalloved.*, **26(2)**, 374.

145. Lucas, L.-D. (1959) Densite d'alliages fer–carbon a l'etat liquide, *Comp. rend. Acad. Sci.*, **248(16)**, 2336.

146. Uberlacker, E. and Lucas, L.-D. (1962) Densite de l'etat du zinc et des alliages etain–zinc a l'etat liquide, *Comp. rend. Acad. Sci.*, **254(9)**, 1622.

147. Haldon, F.A. and Kingery, W.D. (1955) Influence of C, N, O and S on the surface tension of liquid iron, *J. Phys. Chem.*, **59(6)**, 557.

148. Van Tszin-tan, Karasev, R.A., and Samarin, A.M. (1960) Influence of carbon and oxygen on the surface tension of liquid iron, *Izv. Akad. Nauk SSSR. Metallurgiya i Toplivo*, No. 1, 30.

149. Korol'kov, A.M. and Shashkov, D.P. (1962) Electrical resistivity of some alloys in liquid state, *Izv. Akad. Nauk SSSR. Metallurgiya i Toplivo*, No. 1, 84.

150. Toye, T.C. and Jones, E.E. (1958) Physical properties of certain liquid binary alloys of tin and zinc, *Proc. Phys. Soc.*, **71(1)**, 88.

151. Hultgren, R., Orr, R., Anderson, P., and Relley, K. (1963) *Selected values of thermodynamic properties of metals and alloys*, New York– London: J.Wiley and Sons.

152. Bunin, K.P. (1946) On the structure of metallic eutectic melts, *Izv. Akad. Nauk SSSR, OTN*, No. 2, 305.

153. Vertman, A.A., Samarin, A.M., and Yakobson A.M. (1960) On the structure of liquid eutectics, *Izv. Akad. Nauk SSSR, Metallurgiya i Toplivo*, No. 3, 17.

154. Kumar, R. and Sivaramakrishnan, C.S. (1969) Stability of liquid Pb– Cd systems,

J. Mater. Sci., **4(5)**, 377.

155. Kumar, R. and Sivaramakrishnan, C.S. (1969) Structure and stability of Pb–Sb liquid alloys, *J. Mater. Sci.*, **4(5)**, 383.

156. Kumar, R. and Sivaramakrishnan, C.S. (1969) Structure and stability of liquid aluminium–zinc alloys, *J. Mater. Sci.*, **4(11)**, 1008.

157. Izmailov, V.A. and Vertman, A.A. (1971) On the state of silicon in silumin, *Izv. Akad. Nauk SSSR, Metally*, No. 6, 217.

158. Roshchina, G.P. and Ishchenko, E.D. (1964) On the structure of liquid systems with eutectic phase diagram, *Ukr. Fiz. Zh.*, **9(3)**, 334.

159. Nikonova, V.V. and Bartenev, G.M. (1961) Some features of phase diagrams of eutectic-type binary system in relation with the structure of liquid eutectics, *Izv. Akad. Nauk SSSR, Metallurgiya i Toplivo*, No. 3, 131.

160. Klyachko, Yu.A. (1960) On the macromolecular structure of liquid metals and interaction of macromolecules, *Izv. Akad. Nauk SSSR, Metallurgiya i Toplivo*, No. 6, 85.

161. Grigorovich, V.K. (1961) Structure of liquid alloys in relation with phase diagrams, *Izv. Akad. Nauk SSSR, Metallurgiya i Toplivo*, No. 3, 124.

162. Vertman, A.A., Samarin, A.M., and Turovskii, B.M. (1960) The structure of liquid alloys of the iron–carbon system, *Izv. Akad. Nauk SSSR, Metallurgiya i Toplivo*, No. 6, 123.

163. Zalkin, V.M. (1987) *Priroda evtekticheskikh splavov i effect kontaktnogo plavleniya* (The Nature of Eutectic Alloys and Effect of Contact Melting), Moscow: Metallurgiya.

164. Zaiss, W., Steeb, S., and Bauer, G. (1976) Structure of molten Bi–Cu alloys by means of cold neutron scattering in the region of small momentum transfer, *Phys. Chem. Liq.*, **6(1)**, 21.

165. Bellisent-Funel, M.-C., Roth, M., and Desre, P. (1979) Small-angle neutron scattering on liquid Ag–Ge alloys, *J. Phys. F: Metal Phys.*, **9(6)**, 997.

166. Huijben, M.J., Van Lugt, W., and Reimert, W.A. (1979) Investigation on the structure of liquid Na–Cs alloys, *Physica B+C*, **97(4)**, 338.

167. Nikitin, V.I. (1995) *Nasledstvennost' v litykh splavakh* (Heredity in cast alloys), Samara: Gos. Tech. Univ.

168. Baum, B.A., Khasin, G.A., Tyagunov, G.V., *et al.* (1984) *Zhidkaya stal'* (Liquid Steel), Moscow: Metallurgiya.

169. Popel, P.S., Chikowa, O.A., and Matveev, V.M. (1995) Metastable colloidal states of liquid metallic solutions, *High Temperature Materials and Processes*, **4(4)**, 219.

170. Belashchenko, D.K. (1957) Viscous and electrical properties of liquid binary alloys and their relation to the structure of liquid, *Zh. Fiz. Khim.*, **117(1)**, 98.

171. Gotgil'f, T.L. and Lyubimov, A.P. (1965) Investigation of viscosity hysteresis phenomenon in melts of the thallium–bismuth system, *Izv. Vyssh. Ucnebn. Zaved., Tsvet. Met.*, No. 6, 128.

172. Gotgil'f, T.L. and Lyubimov, A.P. (1966) On the structural changes in liquid thallium, *Dokl. Akad. Nauk SSSR, Khimiya*, **170(5)**, 1126.

173. Neimark, V.E. (1961) On the relation between the structure of short-range order of atoms in liquid and the structure of the same substance in the solid state, in: *Stroenie i svoistva zhidkikh metallov* (Structure and Properties of Liquid Metals), Moscow: Fizmatgiz.

174. Chipman, J. (1962) Incomplete mixing in the deoxidation of steel, *Trans. Metallurg. Soc. AIME*, **224(6)**, 1288.

175. Popel, P.S., Baum, B.A., and Kosilov, N.S. (1982) Interphase phenomena in mixing

metallic melts, in: *Adgeziya rasplavov i paika materialov* (Adhesion of Melts and Soldering of Materials), Kiev: Naukova Dumka, No. 9, 8.

176. Spasskii, A.G., Fomin, B.A., and Aleinikov, S.I. (1959) Heat treatment of liquid metals and its influence on mechanical properties of castings, *Liteinoe Proizvodstvo*, No. 10, 35.

177. Novikov, I.I. and Zolotorevskii, V.S. (1966) *Dendritnaya likvatsiya v splavakh* (Dendritic Segregation in Alloys), Moscow: Nauka.

178. Popel, P.S., Presnyakova, E.L., Pavlov, V.A., Arkhangel'skii, E.L., *et al.* (1985) On the origin of microlamination of eutectic Sn–Pb alloys in the liquid state, *Izv Akad. Nauk SSSR, Metally*, No. 2, 53.

179. Gavrilin, I.V. (1985) Sedimentation experiment in studies of liquid alloys, *Izv Akad. Nauk SSSR. Metally*, No. 2, 66.

180. Gol'tyakov, B.P., Popel, P.S., Prokhorenko, V.Ya., and Sidorov, V.E. (1988) Magnetic effects showing the hereditary microheterogeneity of Au–Co melts, *Melts*, 2(6), 83.

181. Popel, P.S., Manov, V.P., and Manukhin, A.B. (1985) Effect of the state of the melt on the structure of Sn–Pb films after solidification, *Dokl. Akad. Nauk SSSR*, 281(1), 107.

182. Popel, P.S., Presnyakova, E.L., Pavlov, V.A., and Arkhangel'skii, E.L. (1985) The region of existence of the metastable quasieutectic structure in the Sn–Pb system, *Izv Akad. Nauk SSSR. Metally*, No. 4, 198.

183. Popel, P.S., Demina, E.L., Arkhangel'skii, E.L., *et al.* (1987) Density and electrical resistivity of Sn–Pb melts in homogeneous and microlaminated states, *Izv Akad. Nauk SSSR, Metally*, No. 3, 52.

184. Makeev, V.V. and Popel, P.S., (1989) An investigation of Fe–Cr and Ni–Cr melts by the method of gamma-ray densitometry, *Zh. Fiz. Khim.*, 62(12), 3278.

185. Makeev, V.V. and Popel, P.S., (1990) Volume characteristics of alloys of the Ni–B system in the region from 1100 to 2170 K, *Zh. Fiz. Khim.*, 64, 568.

186. Popel, P.S. and Sidorov, V.E. (1997) Microheterogeneity of liquid metallic solutions and its influence on the structure and properties of rapidly quenched alloys, *Mater. Sci. Eng.*, A226–228, 237.

187. Kolotukhin, E.V., Popel, P.S., Tyagunov, G.V., *et al.* (1988) Electrical resistivity and density of liquid iron–boron alloys, *Izv. Vyssh. Uchebn. Zaved., Chern. Met.*, No. 6, 68.

188. Kolotukhin, E.V., Popel, P.S., and Tsepelev, V.S. (1988) Electrical resistance of melts of the Co–B system, *Melts*, 2(3), 25.

189. Pushl, W. and Abauer, H.P. (1980), The kinetics of stable heterogeneous binary systems, *Phys. stat. sol. (b)*, 102(2), 447.

190. Zentko, A., Duhaj, P., Potocky, L. *et al.* (1975) Low field magnetic susceptibility of amorphous $Co_xPd_{80-x}Si_{20}$ and $Fe_xPd_{80-x}Si_{20}$ alloys, *Phys. stat. sol. (a)*, 31(1), K41.

191. Butvin, P. and Duhaj, P. (1976) The Hall effect in PdSi based amorphous alloys containing Co, *Czechosl. J. Phys. B*, 26(4), 469.

192. Takahashi, M., Kim Chong Oh, Koshimura, M., and Suzuki, T. (1978) Temperature dependence of saturation magnetization in amorphous Co–B alloys, *Jap. J. Appl. Phys.*, 17(10), 1911.

193. Fish, G.E. and Child, H.R. (1981) Studies of chemical homogeneity and magnetic domain walls in Fe-based metallic glasses using small-angle neutron scattering, *J. Appl. Phys.*, 52(3), Part. 2, 1880.

194. Cser, K., Kovacs, I., Lovas, A., *et al.* (1982) Small-angle neutron scattering of Fe–B and Fe–Ni–B metallic glasses, *Nucl. Instrum. Meth.*, 199(1, 2), 301.

195. Child, H.R. (1981) Small-angle neutron scattering from magnetic correlations in

$Fe_{0.7}Al_{0.3}$, *J. Appl. Phys.*, **52(3)**, Part 2, 1732.

196. Yavari, A.R. and Maret, M. (1983) Anisotropie de la diffusion de neutrons aux petits angles dans les verres metalliques, *Compt. Rend. Acad. Sci., Ser. 2*, **296(21)**, 1637.

197. Osamura, K., Shibue, K., Suzuki, R., and Murokami, Y. (1981) SAXS study on the the structure and crystallization of amorphous metallic alloys, *Colloid and Polym. Sci.*, **259(6)**, 677.

198. Khatanova, N.A. and Kamzeeva, E.E. (1984) Microheterogeneity of the structure of iron-based amorphous alloys, *Vestnik MGU, Fizika, Astronomiya*, **25(3)**, 97.

199. Goltz, G. and Kronmuller, H. (1980) Magnetic small-angle scattering of neutrons in amorphous ferromagnets, *Phys. Lett. A*, **77(1)**, 70.

200. Ishinose, H. and Ishida, Y. (1983) High resolution electron microscopic observation of the structure and relaxation phenomenon of $Fe_{40}Ni_{40}P_{14}B_6$ amorphous alloy, *Trans. Jap. Inst. Metals*, **24(6)**, 405.

201. Khlyntsev, V.P., Potapov, L.P., and Kondrat'ev, V.N. (1981) Study of solidification of metallic $Ni_{60}Nb_{40}$ metallic glass, *Fiz. Met. Metalloved.*, **51(6)**, 1227.

202. Schild, K. (1985) Untersuhung der mittelreichweitigen Struktur von rasch abgeschrekten Nickel- und Eisen-Basislegierungen mittels Röntgen- und Neutronenstereuung im Bereich kleiner Impulsüberträge: Dissertation Dokt. Naturwiss. Stuttgart, 1985.

203. Boucher, B., Chieux, P, Convert, P., and Tournarie, M. (1983) Small-angle neutron scattering determination of medium and long-range order in the amorphous metallic alloy $TbCu_{3.54}$, *J. Phys. F: Metal Phys.*, **13(7)**, 1339.

204. Cocco, G., Enzo, S., Antonione, C., *et al.* (1984) A SAXS study of the crystallization behaviour of the $Pd_{72}B_{28}$ glassy alloy, *J. Non-Cryst. Solids*, **68(2–3)**, 237.

205. Zhao, J.G., Cornelison, S.G., Sellmyert, D.J., and Hadjipanayis, G. (1983) Chemical short-range order and phase separation in Pr−Ga−Fe glasses, *J. Non-Cryst. Solids*, **55(2)**, 203.

206. Ratajczak, H., Slaneo, P., Tima, T., *et al.* (1983) Electrical resistivity and Hall effect in thin amorphous VFe films, *Phys. stat. sol. (a)*, **77(2)**, 785.

207. Iskhakov, R.S., Karpenko, M.M., Popov, G.V., and Ovcharov, V.P. (1986) Investigation of characteristics of local anisotropy of Fe−B amorphous alloys, *Fiz. Met. Metalloved.*, **61(2)**, 265.

208. Piller, J. and Haasen, P. (1982) Atom probe field ion microscopy of a FeNiB glass, *Acta Metall.*, **30(1)**, 1.

209. Terauchi, H., Jida, S., Tanabe, K., *et al.* (1983) Heterogeneous structure of amorphous materials, *J. Phys. Soc. Japan*, **52(10)**, 3454.

210. Hermann, H. and Mattern, N. (1986) Analytic approach to the structure of amorphous iron−boron alloys, *J. Phys. F: Metal Phys.*, **16(2)**, 131.

211. Oamura, K., Ochiai, S., and Takayama, S. (1984) Structure and mechanical properties of a $Fe_{90}Zr_{10}$ amorphous alloy, *J. Mater. Sci.*, **19(6)**, 1917.

212. Haasen, P. (1983) Metallic glasses, *J. Non-Cryst. Solids*, **56(1–3)**, 191.

213. Pompe, G., Gaafar, M., Falz, M., and Buttner, P. (1982) Thermal conductivity of amorphous iron alloys at low temperatures, *Acta Phys. Acad. Sci. Hungarie*, **53(3–4)**, 401.

214. Schaafsma, A.S., Snijders, H., Van der Woude, F., *et al.* (1979) Amorphous to crystalline transformation of $Fe_{80}B_{20}$, *Phys. Rev. B: Solid State*, **20(11)**, 4423.

215. Cahn, J.W. and Hilliard, J.E. (1958) Free energy of a non-uniform system. 1. Interfacial free energy, *J. Chem. Phys.*, **28(2)**, 258.

216. Popel, P.S. and Demina, E.L. (1986) Analysis of the process of mutual solution of liquids with limited miscibility, *Zh. Fiz. Khim.*, **60(7)**, 1602.

217. Barboi, V.M., Glazman, Yu.M., and Fuks, G.I. (1970) On the nature of aggregative stability of colloidal solutions. The conditions for existence of thermodynamically equilibrium two-phase disperse systems, *Colloid. Zh.*, **32(3)**, 321.
218. Lyubov, B.Ya. (1969) *Kineticheskaya teoriya fazovykh prevrashchenii* (Kinetic Theory of Phase Transformations), Moscow: Metallurgiya.
219. Rusanov, A.I. (1964) *Usp. Khim.*, **33**, 873.

Chapter 2

1. Belov, A.F., Dobatkin, V.I., and Drits, M.E. (1987) Technical progress in production of light alloys, *Izv. Akad. Nauk SSSR, Metally*, No. 5, 38.
2. Dobatkin, V.I. and Elagin, V.I. (1981) *Granuliruemye alyuminievye splavy* (Granulated Aluminium Alloys), Moscow: Metallurgiya.
3. Grant, N.J. (1970) *Rapid Solidification Processing*, Los Angeles: Claitor Publ. Div.
4. Popel, P.S., Demina, E.L., Arkhangel'skii, E.L., and Baum, B.A. (1987) Irreversible changes of density of Al–Si melts at high temperatures, *Teplofizika Vysokikh Temperatur*, **25(3)**, 487.
5. Korzhavina, O.A., Brodova, I.G., Nikitin, V.I., Popel, P.S., and Polents, I.V. (1991) Viscosity and electrical resistance of Al–Si melts and the influence of their structural state on the structure of cast metal, *Rasplavy*, No. 1, 10.
6. Zinov'ev, V.E. (1984) *Kineticheskie svoistva metallov pri vysokikh temperaturakh. Spravochnik* (Handbook on Kinetic Properties of Metals at High Temperatures), Moscow: Metallurgiya.
7. Korzhavina, O.A., Mokeeva, L.V., Popel, P.S., Shubina, T.B., and Rozhizina, E.V. (1989) On the influence of the structural state of Al–Ge melt on the mutual solubility of the components in the solid phase, *Rasplavy*, No. 6, 106.
8. Korzhavina, O.A. and Popel, P.S. (1989) The region of existence of metastable microheterogeneity in Al–Sn melts, *Zh. Fiz. Khim.*, **63(3)**, 841.
9. Batalin, G.I., Beloborodova, E.A., and Kazimirov, V.P. (1983) *Termodinamika i stroenie zhidkikh splavov na osnove alyuminiya* (Thermodynamics and Structure of Aluminium-Based Liquid Alloys), Moscow: Metallurgiya.
10. Brodova, I.G., Zamyatin, V.M., and Popel, P.S. (1988) Conditions for the formation of metastable phases in solidification of Al–Zr alloys, *Rasplavy*, **2(6)**, 83.
11. Nikitin, V.I. (1995) *Nasledstvennost' v litykh splavakh* (Heredity in Cast Alloys), Samara: Gos. Tech. Univ.
12. Korzhavina, O.A., Popel, P.S., Brodova, I.G., and Polents, I.V. (1990) Irreversible changes of viscosity of Al–Mn melts at high temperatures, *Rasplavy*, No. 6, 23.
13. Bibik, E.E. (1981) *Reologiya dispersnykh sistem* (Rheology of Disperse Systems), Leningrad: LGU.
14. Lyupis, K. (1980) *Khimicheskaya termodinamika materialov* (Chemical Thermodynamics of Materials), Moscow: Metallurgiya.
15. Golubev, S.V., Korzhavina, O.A., Kononenko, V.I., Popel, P.S., Brodova, I.G., Polents, I.V., and Shubina, T.B. (1991) Influence of viscosity and electrical resistance on the structural state of Al–Sc melts and the structure of cast metal, *Izv. Akad. Nauk SSSR, Metally*, No. 1, 46.
16. Brodova, I.G., Polents, I.V., Korzhavina, O.A., Popel, P.S., Korshunov, I.P., and Esin, V.O. (1990) Structural studies of rapidly quenched Al–Sc alloys, *Rasplavy*, No. 5, 73.
17. Popel, P.S., Chikowa, O.A., Brodova, I.G., and Polents, I.V. (1992) Specific features of the structure formation during solidification of Al–In alloys, *Phys. Metals*

Metallography, **74(3)**, 274.

18. Zhukova, L.A. and Popel, P.S. (1982) Electron diffraction study of the structure of melts, *Zh. Fiz. Khim.,* **56(11)**, 2702.

19. Hothler, J. and Stiib, J. (1975) Struktur von Aluminium–Indium– Schelzen mittels Röntgenweitwinkelbeugung, *Z. Naturforsch.,* **30a(6–7)**, 771.

20. Herwig, F. and Hoyer, W. (1994) Viscosity investigations on liquid alloys of the monotectic system Al–In, *Z. Metallkd.,* **85(6)**, 388.

21. Chikowa, O.A., Sukhanova, T.D., Popel, P.S., Brodova, I.G., and Polents, I.V. (1994) Viscosimetric study of molten Al–Pb alloys, in: "Nano-94" (Int. Conf. Stuttgart Univ. 1994. Abstracts), p. 205.

22. Ubbelohde, A.R. (1965) *Melting and Crystal Structure,* Oxford: Clarendon.

23. Singh, M. and Kumar, K. (1966) Structure of liquid aluminium–copper alloys, *Trans. Indian Inst. Metals,* **19**, 117.

24. Cahn, J.W. (1960) *Acta Metall.,* **8**, 554.

25. Tiller, W.A. (1958) *J. Appl. Phys.,* **29**, 611.

26. Chalmers, B. (1965) *Principles of Solidification,* New York: Wiley.

27. Jackson, K.A. (1958) *Liquid Metals and Solidification,* Cleveland: ASM.

28. Mullins, W.W. (1959) *Acta Metall.,* **7**, 786.

29. Flemings, M.C. (1974) *Solidification Processing,* New York: Mc Graw-Hill.

30. Miroshnichenko, I.S. (1982) *Zakalka iz zhidkogo sostoyaniya* (Quenching from Liquid State), Moscow: Metallurgiya.

31. Brodova, I.G., Popel, P.S., Esin, V.O., and Moiseev, A.I. (1988) Morphological peculiarities of the structure and properties of hypereutectic silumins, *Fiz. Met. Metalloved.,* **65(6)**, 1149.

32. Mondolfo, L.F. (1976) *Aluminium Alloys: Structure and Properties,* London: Butterworths.

33. Brodova, I.G., Bashlikov, D.V., Polents, I.V., and Chikowa, O.A. (1997) Influence of heat time melt treatment on the structure and the properties of rapidly solidified aluminium alloys with transition metals, *J. Mater. Sci. Eng.,* **A226–228**, 136.

34. Brodova, I.G., Polents, I.V., and Esin, V.O. (1992) On the formation of the cast structure of supercooled Al–Ti alloys, *Phys. Metals Metallography,* **73(1)**, 63.

35. Brodova, I.G., Polents, I.V., Bashlikov, D.V., Popel, P.S., and Chikowa, O.A. (1995) The mechanism of formation of ultradisperse phases in rapidly solidified aluminium alloys, in: *Nanostructured Materials,* **6(1–4)**, 477.

36. Hort, S., Kitagawa, H., Masutani, T., and Takehara, A. (1977) Structure and phase decomposition of supersaturated Al–Zr solid solution (rapidly solidified), *J. Japan Inst. Light Metals,* **27(3)**, 129.

37. Nes, E. and Billdal, H. (1977) Nonequilibrium solidification of hypereutectic Al–Zr alloys, *Acta Metall.,* **25(9)**, 1031.

38. Ohashi, T. and Ichikawa, R. (1972) A new metastable phase in rapidly solidified Al–Zr alloys, *Metall. Trans.,* **3(8)**, 2300.

39. Chernov, A.A., Givartizov, E.I., Bagdasarov, Kh.S., *et al.* (1980) *Sovremennaya metallografiya* (Modern Metallography), Moscow: Nauka.

40. Horway, G. and Cahn, T.W. (1961) Dendritic and Spheroidal Growth, *Acta Metall.,* No. 9, 695.

41. Borisov, B.T. (1987) *Teoriya dvukhfaznoi zony metallicheskogo slitka* (Theory of the Two-Phase Zone in Metal Ingots), Moscow: Metallurgiya.

42. Esin, V.O. and Tarabaev, L.P. (1985) Generalized criterion for the transition from the layer to the continuous mechanism of crystal growth, *Phys. stat. sol.,* **90**, 425.

43. Zhuravlev, V.A., Savinskii, S.S., Galenko, P.K., and Popov, A.G. (1992) *Voprosy*

sinergetiki (Problems of Sinergetics), Izhevsk: Udmurtskii Univ.

44. Keith, H. and Padden, F. (1963) Phenomenological Theory of Spherulitic Crystallization, *J. Appl. Phys.*, **344(8)**, 2409.

45. Shubnikov, A.V. (1975) On the nucleation forms of spherulites, *Kristallografiya*, **2(5)**, 584.

46. Esin, V.O., Sazonova, V.A., and Zabolotskaya, I.A. (1989) Spherulitic forms of Crystallization in Metals, *Izv. Akad. Nauk SSSR, Metally*, No. 2, 73.

47. Ohashi, T. and Ichikawa, R. (1973) Grain refinement in aluminium– zirconium and aluminium– titanium alloys by metastable phases, *Z. Metallkd.*, **64(7)**, 517.

48. Vatolin, N.A. and Pastukhov, E.A. (1980) *Difraktsionnye issledovaniya stroeniya vysokotemperaturnykh rasplavov* (Diffraction Studies of the Structure of High-Temperature Melts), Moscow: Nauka.

49. Turovskii, B.M. and Ivanova, I.I. (1974) Studies of the temperature dependence of viscosity of molten silicon, *Izv. Akad. Nauk SSSR. Neorg. Materialy*, No. 12, 2108.

50. Baum, B.A. (1979) *Metallicheskie zhidkosti*, Moscow: Nauka.

51. Morokhov, I.D., Trusov, L.I., and Lapovok, V.I. (1984) *Fizicheskie yavleniya v ultradispersnykh sredakh* (Physical Phenomena in Ultradisperse Media), Moscow: Energoatomizdat.

52. Zhukova, L.A., Popel, P.S., and Razikova, N.I. (1986) Short-range order change in amorphous island aluminium films on heating and melting, *Izv. Akad. Nauk SSSR, Neorg. Materials*, **22(10)**, 1651.

53. Taran, Yu.N. and Mazur, V.I. (1978) *Struktura evtekticheskikh splavov* (Structure of Eutectic Alloys) Moscow: Metallurgiya.

54. Elliott, R. (1983) *Eutectic Solidification Processing: Crystalline and Glassy Alloys*, London, Butterworths.

55. Kurz, W. and Sahm, P.R. (1975) *Gerichtet erstarrte eutektische Werkstoffe*, Berlin: Springer.

56. Ivanov, I.I., Zemskaya, V.S., Kubasov, V.K., *et al.* (1979) *Plavlenie, kristallizatsiya i formooobrazovanie v nevesomosti* (Melting, Solidification and Shape Formation in Zero Gravity Conditions), Moscow: Nauka.

57. Brodova, I.G. and Bashlikov, D.V. (1995) The increase in plasticity of silumin, in: *Light Metals, Conf. Proc., TMS*, p. 879.

Chapter 3

1. Gol'dshtein, Ya. and Mizin, V. (1993) *Inokulirovanie zhelezouglerodistykh splavov* (Inoculation of Iron–Carbon Alloys), Moscow: Metallurgiya.

2. Cibula, H. (1949) The mechanism of grain refinement of sand castings on aluminium alloys, *J. Inst. Met.*, **76**, 321.

3. Palatnik, L. and Papirov, I. (1970) *Orientirovannaya kristallizatsiya* (Directed Solidification), Kiev: Tekhnika.

4. Rebinder, P.A. and Likhtman, M.S. (1932) *Issledovaniya v oblasti prikladnoi fizicheskoi khimii poverkhnostnykh yavlenii* (Studies in the Field of Applied Physical Chemistry of Surface Phenomena), Moscow: ONTI.

5. Sivaramarkisdan, C. and Kumar, R. (1987) *Light Metal.*, **45(9–10)**, 30.

6. Dankov, V.I. (1946) *Fizicheskaya Khimiya* (Physical Chemistry), Moscow: Metallurgiya.

7. Malt'sev, M.V. (1964) *Modifitsirovanie struktury metallov i splavov* (Modification of the Structure of Metals and Alloys), Moscow: Metallurgiya.

8. Rembout, H., *Alum. Technol. 86* Proc. Int. Conf., London (1986), 133.
9. Flemings, M.C. (1974) *Solidification Processing*, New York: Mc. Graw-Hill.
10. Vertman, A.A. and Samarin, A.M. (1969) *Svoistva rasplavov zheleza* (Properties of Iron Melts), Moscow: Nauka.
11. Ershov, G.S. and Chernyakov, V.A. (1978) *Stroenie i svoistva zhidkikh i tverdykh metallov* (Structure and Properties of Liquid and Solid Metals), Moscow: Metallurgiya.
12. Boom, E. (1972) *Priroda modifitsirovaniya splavov tipa siluminov* (The Nature of Modification of Silumin-type Alloys), Moscow: Metallurgiya.
13. Brodova, I.G., Polents, I.V., and Popel, P.S. (1993) The effect of master alloy structure upon grain refinement of aluminium alloys with zirconium, *Phys. Metals Metallography*, **76(5)**, 508.
14. Bazin, Yu.A., Popel, P.S., Domashnikov, B.P., *et al.* (1984) On the possible causes of modification and strengthening of aluminium by scandium, *Izv. Vyssh. Uchebn. Zaved., Tsvet. Met.*, No. 5, 101.
15. Gubenko, A.Ya. (1986) Effect of impurities on bulk and surface properties of liquid alloys, *Izv. Akad. Nauk SSSR, Metally*, No. 3, 25.
16. Vatolin, N.A. (1983) Metallic melts. The state of the art, *Vestnik Akad. Nauk SSSR*, No. 8, 62.
17. Baum, B.A. (1979) *Metallicheskie zhidkosti* (Metallic Liquids), Moscow: Nauka.
18. Lamikhov, L.K. and Samsonov, G.V. (1964) On the modification of aluminium and AL7 alloy with transition metals, *Tsvet. Metally*, No. 8, 77.
19. Chernov, V.V. and Busol, F.I. (1975) On the mechanism of modification of metals, *Izv. Akad. Nauk SSSR, Metally*, No. 2, 71.
20. Brodova, I.G., Zamyatin, V.M., Popel, P.S., Esin, V.O., Baum, B.A., Moiseev, A.I., Korshunov, I.P., and Polents, I.V. (1988) The conditions of formation of metastable phases in solidification of Al–Zr melts, *Rasplavy*, **2(6)**, 23.
21. Brodova, I.G., Polents, I.V., Korzhavina, O.A., Popel, P.S., Korshunov, I.P., and Esin, V.O. (1990) Structural investigations of rapidly solidified Al–Sc alloys, *Rasplavy*, No. 5, 73.
22. Brodova, I.G., Polents, I.V., and Esin, V.O. (1992) On the formation of the cast structure of supercooled Al–Ti alloys, *Phys. Metals Metallography*, **73(1)**, 63.
23. Polents, I.V., Brodova, I.G., Bashlykov, D.V., *et al.* (1995) The role of kinetics of solution of intermetallic compounds in alloying aluminium melts with titanium, *Rasplavy*, No. 6, 29.
24. Popel, P.S., Zamyatin, V.M., Domashnikov, B.P., *et al.* (1983) Influence of scandium on properties of liquid aluminium, *Izv. Akad. Nauk SSSR. Metally*, No. 3, 38.
25. Nikitin, V.I. (1995) *Nasledstvennost' v litykh splavakh* (Heredity in Cast Alloys), Samara: Gos. Tech. Univ.
26. Elagin, V.I. (1978) *Legirovanie deformiruemykh alyuminievykh splavov perekhodnymi metallami* (Alloying of Deformed Aluminium Alloys with Transition Metals), Moscow: Metallurgiya.
27. Dobatkin, V.I. and Malinovskii, R.R. (1971) Methods of grain refinement of primary crystals of intermetallic compounds in castings of aluminium alloys, in: *Struktura i svoistva legkikh splavov* (Structure and Properties of Light Alloys), Moscow: Nauka, p. 82.
28. Dobatkin, V.I. and Elagin, V.I. (1981) *Granuliruemye alyuminievye splavy* (Granulated Aluminium Alloys), Moscow: Metallurgiya.
29. Eskin, G.I. (1988) *Ul'trazvukovaya obrabotka rasplavlennogo alyuminiya* (Ultrasonic Treatment of Molten Aluminium), Moscow: Metallurgiya.

30. Mondolfo, L.F. (1976) *Aluminium Alloys: Structure and Properties*, London: Butterworths.
31. Novikov, I.I. (1978) *Teoriya termicheskoi obrabotki metallov* (Theory of Heat Treatment of Metals), Moscow: Metallurgiya.
32. Semenchenko, V.K. (1957) *Poverkhnostnye yavleniya v metallakh i splavakh* (Surface Phenomena in Metals and Alloys), Moscow: Gostekhizdat.
33. Efimenko, V.N., Baranov, A.D., Kisun'ko, V.Z., and Bychkov, Yu.B. (1982) On the modifying effect of certain microadditions on the solidification of silumin, *Izv. Vyssh. Uchebn. Zaved., Tsvet. Met.*, No. 6, 86.
34. Mazur, V.I., Osetrov, S.A., and Prigunova, A.G. (1977) A model of the structure of eutectic melts Al–Si, Al–Si–Na and Al–Cu, *Fiz. Met. Metalloved.*, **43**, 1021.
35. Brodova, I.G., Bashlykov, D.V., Polents, I.V., and Yablonskikh, T.I. (1995) Effect of small tin additions on the structure of hypereutectic silumin, *Phys. Metals Metallography*, **79(4)**, 425.
36. Mazur, V.I., Prigunova, A.G., and Taran, Yu.N. (1980) Models of melts in the Al–Si system from the results of structural analysis of products of quenching from the liquid state, *Fiz. Met. Metalloved.*, **50(1)**, 123.
37. Lamikhov, L.K. and Samsonov, G.V. (1963) On the modification of aluminium with transition metals, *Izv. Akad. Nauk SSSR, OTN, Metallurgiya i gornoe delo*, No. 3, 96.
38. Emel'yanov, A.V., Bazin, Yu.A., and Baum, Yu.A. (1987) Effect of additions of 3d-transition metals on the structural characteristics of liquid aluminium, *Izv. Vyssh. Uchebn. Zaved., Fizika*, No. 7, 96.
39. Popel, P.S., Chikova, O.A., and Matveev, V.M. (1995) Metastable colloidal states of liquid metallic solutions, *J. High Temperature Materials and Processes*, No. 4, 219.
39. Korzhavina, O.A. and Popel, P.S. (1989) The region of existence of metastable microheterogeneity in Al–Sn melts, *Zh. Fiz. Khim.*, **63(3)**, 841.
40. Belyaev, A.I and Wol'fson, G.E. (1967) *Poluchenie chistogo aluminiya* (Pure Aluminium Production), Moscow: Metallurgiya.
41. Napalkov, V.I., Bondarev, B.I., *et al.* (1983) *Ligatury dlya proizvodstva alyuminievykh i magnievykh splavov* (Master Alloys for Making Aluminium and Magnesium Alloys), Moscow: Metallurgiya.
42. Chikowa, O.A., Sukhanova, T.D., Popel, P.S., Brodova, I.G., and Polents, I.V. (1994) Viscosimetric study of molten Al–Pb alloys, in: "Nano-94" (Int. Conf., Stuttgart Univ. 1994. Abstracts), p. 205.
43. Lang, G. (1974) *Aluminium*, **50(11)**, 731.
44. Fridlyander, I.N., Drits, A.M., and Eliseev, A.A. (1986) Effect of scandium on the structure and properties of alloys of the Al–Cu system, in: *Legkie i zharoprochnye splavy i ikh obrabotka* (Light and Heat-resistant Alloys and Their Treatment), Moscow: Nauka, p. 126.
45. Korzhavina, O.A., Brodova, I.G., Nikitin, V.I., Popel, P.S., and Polents, I.V. (1991) Viscosity and electrical resistance of Al–Si melts and the influence of their structural state on the structure of cast metal, *Rasplavy*, No. 1, 10.
46. Nizhenko, V.I. and Floka, L.I. (1987) *Poverkhnostnoe natyazhenie zhidkikh metallov i splavov* (Surface Tension of Liquid Metals and Alloys), Moscow: Metallurgiya.
47. Stroganov, G.B. (1975) *Vysokoprochnye liteinye alyuminievye splavy* (High-Strength Casting Aluminium Alloys), Moscow: Metallurgiya.
48. *Aluminum: Properties and Physical Metallurgy* (1984), Ed. Hatch, J.E., Metals Park, Ohio: ASM.
49. Chalmers, B. (1965) *Principles of Solidification*, New York: Wiley.
50. Brodova, I.G., Bashlikov, D.V., Polents, I.V., and Chikowa, O.A. (1997) Influence of

heat time melt treatment on the structure and the properties of rapidly solidified aluminium alloys with transition metals, *J. Mater. Sci. Eng.*, **A226–228**, 136.

Chapter 4

1. Frenkel, Ya.I. (1959) *Kineticheskaya teoriya zhidkosti* (Kinetic Theory of Liquids), Moscow: Izd. Akad. Nauk.
2. Eskin, G.I. (1988) *Ultrazvukovaya obrabotka rasplavlennogo alyuminiya* (Ultrasonic Treatment of Molten Aluminium), 2nd edition, Moscow: Metallurgiya.
3. Eskin, G.I. (1996) Degassing, filtration and grain refinement processes of light alloys in a field of acoustic cavitation, in: *Advances in Sonochemistry*, ed. T. J. Mason, London: JAI Press, **4**, 101.
4. Eskin, G.I. (1998) Ultrasonic Treatment of Light Alloys Melts, Amsterdam: Gordon & Breach.
5. Sirotyuk, M.G. (1974) Experimental study of ultrasonic cavitation, in: *Moshchnye ul'trazvukovye polya* (Powerful Ultrasound Fields), ed. L. D. Rozenberg, Moscow: Nauka, p. 167.
6. Harvey, E.W., Whitely, A.N., McElroy, W.D., *et al.* (1945) *Trans. Am. Chem. Soc.*, **67**, 152.
7. Popel, P.S., Pavlov, V.A., Konovalov, V.A. *et al.* (1985) *Tsvet. Met.*, No. 7, 12.
8. Popel, P.S. and Eskin, G.I. (1996) Effect of ultrasonic treatment on microheterogeneity of liquid alloys, in: *Proc. 14th Int. Conf. on Utilization of Ultrasonic Methods in Condensed Matter*, part 1, 127, Slovakia.
9. Eskin, G.I. (1994) *Ultrasonic Sonochemistry*, **1(1)**, 59.
10. Popel, P.S. and Kuzin, S.N. (1997) Gas-saturated metallic melts as nanodispersed foams, in: *High Temperature Capillarity*, 2nd Int. Conf., Krakow, Poland, 1997, Abstracts, 56.
11. Flemings, M.C. (1974) *Solidification Processes*, New York: McGraw-Hill.
12. Elagin, V.L., Zakharov, V.V., and Rostova, T.D. (1982) *Tsvet. Met.*, No. 2, 96.
13. Kattamis, T.Z., Holmberg, U.T., and Flemings, M.C. (1967) *J. Inst. Met.*, **95(11)**, 343.
14. Patterson, R.I., Cox, A.R., and Van Reath, E.C. (1980) *J. Met.*, **32(9)**, 34.
15. Dobatkin, V.I. and Eskin, G.I. (1991) *Tsvet. Met.*, No. 12, 64.
16. Dobatkin, V.I., Belov, A.F., Eskin, G.L., Borovikova, S.I., and Gol'der, Yu.G. (1984) Scientific discovery "Regularity of crystallization of metal materials", No. 271, *Vestnik Akad. Nauk SSSR*, No. 1, 137.
17. Flemings, M.C. (1991) *Metall. Trans.*, **22B(6)**, 269.
18. Dobatkin, V.I. and Eskin, G.I. (1996) *Proc. 4th Int. Conf. SSM*, Sheffield, UK, 133.
19. Eskin, G.I., Pimenov, Yu.P., and Makarov, G.S. (1996) *Proc. 5th ICAA*, Grenoble, France (Mater. Sci., Forum, 1997, **242**, 65).
20. Dobatkin, V.I. (1960) *Slitki alyuminievykh splavov* (Ingots of Aluminium Alloys), Moscow: Metallurgizdat.
21. Dobatkin, V.I. and Elagin, V.I. (1980) *Granuliruemye Alyuminievye Splavy* (Granulated Aluminium Alloys), Moscow: Metallurgiya.
22. Bondarev, B.I., Napalkov, V.I., and Tararushkin, V.I. (1978) *Modifitsirovanie deformiruemykh alyuminievykh splavov* (Modification of Deformed Aluminium Alloys), Moscow: Metallurgiya.

INDEX

Milton Keynes UK
Ingram Content Group UK Ltd.
UKHW051949071024
449327UK00026B/2233